Aerodrome Control

机 场 管 制

王同乐　张飞　姜涛　编

中国民航出版社

图书在版编目（CIP）数据

机场管制 = Aerodrome Control：英文/王同乐，
张飞，姜涛编 . —北京：中国民航出版社，2011. 9
ISBN 978-7-5128-0048-9

Ⅰ.①机…　Ⅱ.①王… ②张… ③姜…　Ⅲ.①机场-
空中交通管制-教材-英文　Ⅳ.①V355. 1

中国版本图书馆 CIP 数据核字（2011）第 175587 号

责任编辑：杜文晔

机 场 管 制

王同乐　张飞　姜涛　编

出版	中国民航出版社
地址	北京市朝阳区光熙门北里甲 31 号楼 （100028）
排版	中国民航出版社照排室
印刷	中国电影出版社印刷厂
发行	中国民航出版社 （010）64297307　64290477
开本	787×1092　1/16
印张	16
字数	361 千字
版本	2012 年 3 月第 1 版　2012 年 3 月第 1 次印刷
书号	ISBN 978-7-5128-0048-9
定价	34. 00 元

（如有印装错误，本社负责调换）

Preface

Aerodrome control service is part of Air Traffic Service which is provided for aerodrome traffic to ensure the separation and efficient movement of aircraft and vehicles operating on the taxiways and runways of the aerodrome itself, and aircraft in the vicinity of the aerodrome as well. Currently in many countries, aerodrome control service has been regarded as the basic training programme for a new recruiting air traffic controller.

To meet the requirement of rapid developing national civil aviation industry, CAUC decided to set up "Aerodrome Control" as the major course for ATC subject. This book can be used as teaching material for academy students major in ATM who have basic ATM theory and correspondent aviation English capability, as well as a guidance material for OJT trainees from current ATC working unit. The book is written in English, that will be very helpful for trainee's language proficiency, and includes 9 chapters and 3 appendixes. Chapter 1 begins with general introduction of an aerodrome, Chapters 2 – 4 introduce aerodrome marking and signs, aeronautical ground lighting and visual signals, essential aerodrome information which is very essential to aerodrome control tower. Chapters 5 – 7 present separation in the vicinity of aerodrome, and aerodrome control service procedure including standard telephony communication. Chapter 8 gives an example of local instruction which will be used for simulation practice and chapter 9 introduces the guidance and methods to handle emergency situations. Exercises and consulting questions have been given at the end of each chapter. These questions and exercises are very helpful for trainees to understand and master concept and knowledge in each chapter. On material phraseology chosen, the book includes all terms pertain to aerodrome control in reference with English proficiency (ICAO Doc9835). This teaching material consists of two parts which are theory principle and simulation practice with exercises at the end of each chapter to facilitate teaching organization for instructor and self – study of the trainees. By using the book, trainees can acquaint and understand the aerodrome control fundamental theory which includes aerodrome operation, aerodrome markings and signs, provision of aerodrome service, separation, the principle and common used method for emergency handling together with standard telephony communication, etc. In practice part, it enables trainee to achieve the requirement of basic aerodrome control skills by practice using ATC simulation training system which was developed and owned by CAUC.

The editors, Mr. Wang tongle, Zhang fei, Jiang tao summed up the previous teaching experience, took into account current operation and requirements of the ATC unit, assimilated and referred aerodrome control training experience and advantages of the countries including the United States, Australia and Eurocontrol.

For further improvement and betterment, the editors sincerely appreciate precious advices from all readers.

We would like to acknowledge the contributions that were made by the following people: Mr. Yan Shaohua, Deputy Dean of ATM academy who gave valuable opinions and support contributing to the successful completion of this book. Mr. Liuxin, Mr. Zhoujian and Mr. Zhonghan, instructor of ATM academy who gave us their comments and editorial advice, and all experts from regional ATMBs for their reviewing, precious comments and advice. Finally we wish to gratefully acknowledge CAUC and ATMB of CAAC for the opportunity that made this book published.

Table of Contents

Chapter 1

Introduction to aerodrome

A flight normally begins and ends at an aerodrome. Aerodrome is a defined area on land or water (including any buildings, installations, and equipment) intended to be used either wholly or in part for the arrival, departure and surface movement of an aircraft. It varies from a small sod field served for general aviation to a large complex one utilized by air carriers. Aerodrome physical characteristics refer to description, reference code and physical data about orientation, length, width, strength and slope of runway, stopway, taxiway, apron at an aerodrome.

The purpose of this chapter is to provide fundamental and necessary knowledge of an aerodrome to preliminary air traffic controller students and trainees. Through the study, students or trainees shall establish the basic concepts of an aerodrome functions, know the aerodrome physical characteristics such as reference code and major components of airfield area, comprehend the basic principles for institution of aerodrome operation standard minima.

1.1   Types of aerodrome

In air traffic control (ATC) aspects, there are two types of aerodromes, controlled aerodromes and uncontrolled aerodromes.

A controlled aerodrome has an operating control tower. Air Traffic Control (ATC) is responsible for providing for the safe, orderly, and expeditious flow of air traffic at aerodromes where flight operation or traffic volume requires such a service. Pilots operating at a controlled aerodrome need maintain two-way radio communication with aerodrome controllers, and to acknowledge and comply with their instructions. Pilots advise ATC when they cannot comply with the instructions and request amended instructions. When pilot deviates from air traffic instruction in an emergency situation, they need advise ATC of the deviation as soon as possible.

An uncontrolled aerodrome does not have an operating control tower. Two-way radio communications are not required, although it is a good operating practice for pilots to transmit their intentions on the specified frequency for the benefit of other traffic in the area. In China, only some of small airfields which serve to general aviation do not have air traffic control tower.

1. 2　Aerodrome reference code

An aerodrome reference code which is composed of two elements related to the aircraft performance characteristics and dimensions is selected for aerodrome planning purposes. Element 1 is a number (1-4) based on the aircraft reference field length and element 2 is a letter (A-F) based on the aircraft wing span and outer main gear wheel span (see Table 1-1).

Table 1-1　Aerodrome reference code (meters)

Code element 1		Code element 2		
Number	Reference Field length	Letter	Wing span	Gear wheel span
1	<800	A	<15 <4. 5	
2	800 ~ <1200	B	15 ~ <24	4. 5 ~ <6
3	1200 ~ <1800	C	24 ~ <36	6 ~ <9
4	≥1800	D	36 ~ <52	9 ~ <14
		E	52 ~ <65	9 ~ <14
		F	65 ~ <80	14 ~ <16

The intent of the reference code is to provide a simple method for interrelating the numerous specifications concerning the characteristics of aerodromes so as to provide a series of aerodrome facilities that are suitable for the aircrafts intended to operate at the aerodrome. The code is not intended to be used for determining runway length or pavement strength requirements. The typical aircraft type for reference code is as Table 1-2.

Table 1-2　Typical aircraft type corresponding to aerodrome reference code

Code	Typical aircraft type	Code	Typical aircraft type
4F	A380	3A	SAAB
4E	B747	2A	Cessna
4D	A300	1A	Y-5
4C	B737		

1. 3　Certification of aerodromes

A certified aerodrome refers to the aerodrome whose operator has been granted an aerodrome certificate issued by the appropriate authority under applicable regulations for its operation. Certified aerodrome has landing and take-off field, facilities, services (including air traffic control service), equipment, operating procedures, organization and management including a safety management system. In China, the aerodrome certificate is issued by Civil Aviation Administration of China.

1. 4　Airfield area

Airfield area refers to a range of space prepared for the take-off, landing, taxing or parking of an aircraft. It consists of manoeurving area and movement area which includes runway strips, runway end safety area, taxiway, apron and even aerodrome obstacle free space. Manoeuvring area means the part of an aerodrome to be used for the take-off, landing and taxiing of aircraft excluding aprons. Movement area consists of manoeuvring area and the apron (s) and defined to be used for the take-off, landing and taxiing of an aircraft (see Fig 1. 1).

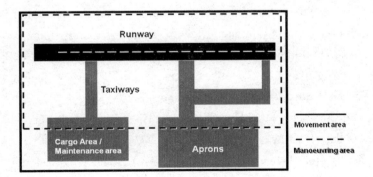

Fig 1. 1　Airport manoeuving area and movement area

1. 4. 1　Runway

Runway is a defined rectangular area on a land aerodrome prepared for the landing and take-off of aircraft. The surface of a runway is constructed without irregularities that would not result in loss of friction characteristics or affect the take-off or landing even when the runway is wet. The strength of a runway is capable of withstanding the traffic of aircraft the runway is intended to serve.

Many factors may effect the determination of location and orientation of a new runway. One important factor is the usability factor, which is determined by the wind distribution and the other important factor is the alignment of the runway to facilitate the provision of approaches conforming to the approach surface specifications. Generally, the number and orientation of runways at an aerodrome is such that the usability factor of the aerodrome is not less than 95 percent for the aircraft that the aerodrome is intended to serve.

The actual runway length to be provided for runway definitely is adequate to meet the operational requirements of the aircraft for which the runway is intended to serve and not less than the longest length determined by applying the corrections for local conditions to the operations and performance characteristics of the relevant aircraft. Both take-off and landing requirements need to be considered when determining the length of runway and the need for operations to be conducted in both directions of the runway. Width of runway is relevant and determined by the aerodrome reference code. The longitudinal slope computed by dividing the difference between the maximum and minimum elevation along the runway centre line by the runway length, basically will not exceed 1.5 per cent where the code number is 3 or 4 and 2 percent where the code number is 1 or 2.

1.4.2 Declared runway distances

The runway distances is calculated to the nearest meter or foot which are very important information for each runway intended for use by international commercial air transport for safety operation. The following distances will be published in AIP (see Fig 1.2).

(a) Take-off run available (TORA). The length of runway declared available and suitable for the ground run of an aeroplane taking off.

(b) Take-off distance available (TODA). The length of the take-off run available plus the length of the clearway, if provided.

(c) Accelerate-stop distance available (ASDA). The length of the take-off run available plus the length of the stopway, if provided.

(d) Landing distance available (LDA). The length of runway which is declared available and suitable for the ground run of an aeroplane landing.

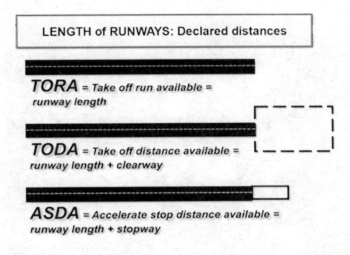

Fig 1. 2   Runway declared distances

1. 4. 3   Runway strips

Runway strips is a defined area including the runway and stopway. The purpose of runways strips (if provided) is to reduce the risk of damage to aircraft running off a runway and to protect aircraft flying over it during take-off or landing operations. The strip shall extend before the threshold and beyond the end of the runway or stopway for certain of distances.

1. 4. 4   Clearway and stopway

Clearway is a defined rectangular area on the ground or water under the control of the aerodrome authority, selected or prepared as a suitable area over which an aircraft may make a portion of its initial climb to a specified height. Stopway is a defined rectangular area on the ground at the end of take-off run available prepared as a suitable area in which an aircraft can be stopped in the case of an aborted take-off (see Fig 1. 3). The decision to provide a stopway and clearway as alternative to an increased length of runway will depend on the physical characteristics of the area beyond the runway end, and on the operating performance requirements of the prospective aircraft. The runway, stopway and clearway lengths to be provided are determined by the aircraft take-off performance, but a check should also be made of the landing distance required by the aircraft using the runway to ensure that adequate runway length is provided for landing. The length of a clearway, however, cannot exceed half the length of take-off run available.

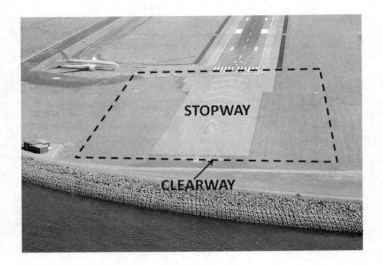

Fig 1. 3   Stopway and clearway

1. 4. 5   Runway end safety area （RESA）

Runway end safety area is an area symmetrical about the extended runway centre line and adjacent to the end of the strip primarily intended to reduce the risk of damage to an aircraft undershooting or overrunning the runway. Runway end safety area generally extends from the end of a runway strip to a distance of at least 90 m.

1. 4. 6   Taxiway

Taxiway is a defined path on a land aerodrome established for the taxiing of aircraft and intended to provide a link between one part of the aerodrome and another. Taxiways should be provided to permit the safe and expeditious surface movement of aircraft. Sufficient entrance and exit taxiways for a runway are also be provided to expedite the movement of aircraft to and from the runway and provision of rapid exit taxiways considered when traffic volumes are high. To reduce the risk of runway incursions, the number of taxiway entrances having direct access to a runway meets the minimum required for efficient runway use.

1.4.7 Apron

Apron is a part of an aerodrome, other than the maneuvering area, intended to accommodate the loading and unloading of passengers and cargo, the refueling, servicing, maintenance and parking of aircraft, and any movement of aircraft, vehicles and pedestrians necessary for such purposes. Aprons generally are provided where necessary to permit the on-and-off-loading of passengers, cargo or mail as well as the servicing of aircraft without interfering with the aerodrome traffic. The total apron area should be adequate to permit expeditious handling of the aerodrome traffic at its maximum anticipated density. In additional to that, each part of an apron is capable of withstanding the traffic of the aircraft it is intended to serve, due consideration being given to the fact that some portions of the apron will be subjected to a higher density of traffic and, as a result of slow moving or stationary aircraft, to higher stresses than a runway.

1.5 Aerodrome operation minima

Aerodrome operation minima are a kind of criteria which apply to takeoff or landing limit. For takeoff, it's prescribed by visibility (VIS) or runway visual range (RVR), including ceiling if necessary; for the precision approach landing, it's prescribed by VIS or/and RVR with altitude/height (DA/DH) and VIS, minimum descent altitude/ height (MDA/MDH) together with ceiling for non-precision approach landing.

Aviation operators will take into account the following factors when they confirm their operating aerodrome's operation minimum standard (company standard) sufficiently:

(a) Aircraft type, performance and controlling character;

(b) The form, technical level and flight experience of flight crews;

(c) Size and character or the runway in use;

(d) Performance and satisfactory degree of available visual navaids and radio navigation facilities;

(e) Pilotage and flight control onboard facilities available for approach landing and go around;

(f) Obstacles and instrument approach OCH within approach zone and go around zone;

(g) Facilities used by the aerodrome for weather report;

(h) Obstacles and necessary obstacle clearance margin within the climb zone.

All kind of instrument arrival and departure procedures, takeoff and landing minima will be published by contracted states to operators, in the form of instrument approach charts, aerodrome charts, instrument arrival and departure charts for each aerodrome according to its approach procedure and published obstacle clearance altitude/obstacle clearance height (OCA/OCH) for different aircraft categories in each state's AIP.

1.5.1 Institution of aerodrome takeoff minima

To institute takeoff minima of an aerodrome, the following factors which could impact takeoff performance, will be taken into consideration:

(a) Avoid disadvantage terrain and obstacle;

(b) Aircraft controllability and performance;

(c) Available visual navaids;

(d) Runway characteristics;

(e) Available navigation facilities;

(f) Abnormal conditions as engine failure;

(g) Disadvantage weather conditions such as contaminated runway, side-wind influence, and etc.

Takeoff minima are usually prescribed by visibility, and when obstacles are clearly observed and avoided, it should include visibility, ceiling and the specific positions must be marked in the published departure procedure chart.

The basic standards for takeoff minimum of an aerodrome are as followings:

(a) For airplane with one or two engines, visibility is 1600m;

(b) For airplane with three or four engines, the visibility is 800m;

(c) Minimum ceiling for single engine airplane shall not be lower than 100m, visibility shall not be lower than 1600m;

(d) The ceiling in takeoff minimum standard shall be at least 60m higher than controlled obstacles. The figure of ceiling shall be rounded 10m up;

(e) Visibility shall ensure flight crew observe and avoid obstacles clearly 500m or 5000m plus the shortest distance from the liftoff end of runway to the obstacle, whichever is lower. The minimum visibility shall not be lower than 1500m for airplanes category A, B and 2000m for category C, D.

1.5.2 Minimum standard for non-precision straight-in approach

The minimum standard for non-precision straight-in approach consists of two parts: minimum descent altitude/height (MDA/MDH) and visibility.

The determination of MDA/MDH is based on OCA, which is prescribed by instrument approach procedures. MDA/MDH may be higher but impossible lower than the OCA/ OCH. To take into account various factors as airplane performance, airborne equipment, skill proficiency and experiences of flight crew etc. , the operator may add a margin to determine the calculation of MDA/MDH based on obstacle clearance.

MDH should not be lower than the minimum height for flying safely while applying the following non-precision approach navigation facilities criteria:

ILS glide slope inoperative, 75m;

VOR with FAF, 75m;

VOR without FAF, 90m;

NDB with FAF, 75m;

NDB without FAF, 90m.

The minimum visibility, which enable the pilot descend safely and timely from the MDA/MDH and make a maneuver landing is based on airplane category, MDA/MDH, available visual navigation aids and its approach types (straight approach or circling approach) . Determination of minimum visibility for non-precision approach refers to the minimum descent height and available visual navigation aids.

1.5.3   Institution of minimum standard of visual circling approach

During visual flight phase after completion of instrument approach, pilot will be able to continuously observe the runway threshold, approach lights or other runway identified marks so as to control the airplane flying within the visual circling area, which generally locates at the opposite direction or another runway landing position. Each aerodrome will set down the minimum standard for visual circling approach. Whenever the aerodrome has adequate conditions, visual circling approach with defined path should be made as possible as it could be.

For some aerodromes with high obstacle at one side of runway or limited airspace in vicinity of the aerodrome, circling approach may be constraint to the side with lower terrain or no airspace limit. Under this circumstance, limitation of visual circling flight will be clearly stipulated in the instrument approach chart and note, e. g. "visual circling approach is only allowed at (Direction description) side of runway. "

Minimum standard for visual circling approach shall contain two key factors: MA/MH and visibility. MA/MH for each category of airplane is determined based on obstacle clearance height. Minimum visibility refers to the minimum descent height and available visual navigation aids. Minimum standard for circling approach is not lower than the minimum standard of straight approach in the same aerodrome.

1.5.4   Institution of minimum standard for precision approach CAT I

Minimum standard for precision approach CAT I consists of decision altitude/height (DA/DH), RVR or visibility. For the runway equipped with RVR, the precision approach minima criteria are prescribed by DA/DH, RVR. For the precision approaches which includes ILS, MLS and PAR approach, when the decision height is 60m or higher and meanwhile minimum visibility 800m or above (RVR 550 or above), the approaches are categorized as precision approach CAT I.

DH is calculated on basis of obstacle clearance height and shall not be lower than one of the following values:

(a) Allowable minimum altitude/height of instrument flight for the aircraft, which indicated in the aircraft's Flight Manual;

(b) Reachable minimum altitude/height completely referring to flight instruments by precision approach navigational facilities;

(c) OCH;

(d) DH.

The minimum weather conditions for the pilot to obtain the required visual reference below DA/DH are prescribed as visibility or RVR (RVR is mandatory when visibility below 800m). Required suitable length of visual segment is dependent on the pilot's eye height, downwards cut-off angle and modes of ground visual navigation aids facilities.

1.5.5   Institution of minimum standard for precision approach CAT II

Minimum standard for precision approach CAT II consists of two factors: DH and RVR. Visibility cannot be used. DH is below 60m but no lower than 30m, RVR is below 550m, but no lower than 350m.

DH of CAT II precision approach is based on OCH and not lower than one of the following values:

(a) Minimum DH prescribed in the airplane airworthiness certificate and the minimum heights used by the visual reference are not required by precision approach navigation facilities;

(b) DH which is approved to be used by the crew;

(c) Minimum DH of 30m for CAT II operations.

Minimum RVR required for CAT II precision approach depends on required visual references when the pilot controls the airplane with different modes, and runway requirements are stipulated in relevant regulations.

1.5.6   Institution of minimum standard for precision approach CAT III

CAT III is a kind of all-weather operation which means the approaches and landings is operated following the special CAT III approach procedures under instrumental conditions to CAT III operational minima. CAT III operational minimum standard is: DH lowers than 30m and control RVR lower than 350m. It also includes the operations using Alert Height (AH) which at this time is 30m or below (without DH) and control RVR lower than 350m.

CAT III operations fall into two categories: fail-passive and fail-operational. Fail-passive operation is limited to CAT IIIA weather conditions (DH15m/RVR200m), while fail-operational operation may proceed to CAT IIIA or CAT IIIB weather conditions. In additional, CAT IIIA requires ground and airborne equipment provide continuous precise guidance till touching down and the CAT IIIB requires ground and airborne equipment provide continuous precise guidance till touching down and rolling to a safe speed.

Exercise

I. Explain the following definitions.

(1) movement area   (2) LDA   (3) ASDA   (4) RESA   (5) TORA

(6) MDA   (7) DH   (8) TODA   (9) stopway

II. Answer the following questions.

(1) In ATC aspects, what are the two types of aerodromes and what's the difference between them?

(2) What factors shall be taken into consideration for institution of aerodrome takeoff minima?

(3) What runway declared distances shall be published in AIP? How to calculate ASDA?

III. Mark the following designated area on the aerodrome picture.

(1) movement area　　(2) manueuvring area　　(3) RESA　　(4) Runway strips

(5) clearway　　(6) stopway　　(7) apron　　(8) taxiway　　(9) exit taxiway

(10) TODA

Chapter 2

Aerodrome surface markings and signs

Surface markings and signs are provided on aerodromes in order to assist pilots in identifying certain locations and to provide guidance for ground movement by day. For aerodrome controller, it's necessary to acquaint these visual aids to keep situation awareness and to facilitate aerodrome control service. The purpose of this chapter is to introduce the detail information of aerodrome surface markings and signs, give students fundamental knowledge for the following study.

2.1　Surface markings

2.1.1　Paved surface markings

Paved Surface Markings are normally produced by the application of skid resistant paints or thermo-plastic materials directly onto the pavement. According to the area where they are painted, the markings can be divided into three categories namely Paved Runway Markings, Paved Taxiway Markings and Paved Apron Markings. They carry information which is crucial for flight operation and some of the most important markings are selected to be described in the following paragraphs.

2.1.1.1　Paved runway markings

(a) Runway designation marking

All paved runways in regular use are identified by a Runway Designation Marking. This marking consists of a two digit number indicating the magnetic heading of the runway to the nearest 10 degrees as shown in Fig 2.1. At those aerodromes with parallel runways where the same magnetic heading applies to more than one runway, the designation marking will include a letter, such as 'L' identifying the left runway as seen from the approach, 'C' identifying the centre runway where there are 3 parallel runways or 'R' for the right runway.

Fig 2. 1   Example of runway designation marking

(b) Threshold, edge and centre line markings

All paved runways in regular use have centre line and threshold markings (see Fig 2.2), the latter varying from the runway designation number alone to separate threshold and designation markings, according to the classification of the runway. Runway edge marking is normally provided on all ILS equipped runways and those other runways where there is insufficient contrast between the runway and its shoulders or where the declared runway width is less than the apparent width.

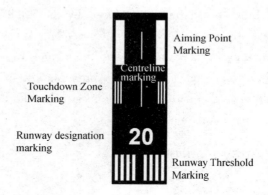

Fig 2. 2   Touchdown zone, aiming point, threshold and centerline markings

Threshold markings are normally located at the beginning of the paved runway surface, and they may be displaced along the runway where, for example, there are obstructions on the approach or where the first portion of the pavement is not fully loading bearing. Where displacement is of a temporary nature – e. g. to accommodate runway maintenance – the normal threshold markings will be obscured and the appropriate displaced threshold marking and threshold marker boards, illustrated in Fig 2.3, put in place in order to mark the new threshold. Whenever a threshold is displaced, the pre-threshold area will be marked according to its bearing strength as illustrated at Fig 2. 4.

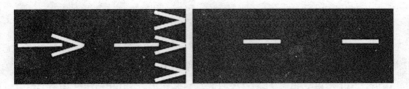

Fig 2. 3   Temporary displaced threshold markings

Fig 2. 4   Displaced threshold marking

(c) Touchdown zone and aiming point

All ILS equipped runways and those other runways where the touchdown zone is insufficiently conspicuous are provided with Touchdown Zone and Aiming Point markings as shown in Fig 2-2. These markings are intended to give added visual texture to the runway surface, particularly in conditions of poor visibility; they also indicate the optimum touchdown zone on the runway. The apparent distance between the Aiming Point marking and the Threshold Marking, as seen from the approach, is intended to aid pilots in judging their angle of approach.

2. 1. 1. 2 Unpaved surface markings

Unpaved Surface Markings are normally confined to unpaved runways and consist of Unpaved Runway Edge, Centre line, Threshold and End Markings.

The edges of unpaved runways are delineated by markers placed at regular intervals along the declared edges of the runway. Where provided, a centre line marking consists of rectangular markers inset flush with the runway surface and spaced at regular intervals along the declared runway centre line. Edge and centre line markers are normally white but may be of any single color that best contrasts with the background.

The threshold and end of an unpaved runway are provided with markers of a similar type, size and color as the edge markers. These markers are placed along the declared threshold and end of the runway and so positioned in relation to the edge markers as to form an 'L' shaped mark at each corner of the runway. In addition, each threshold is marked with a two character designator showing the magnetic heading of the runway to the nearest whole ten degrees.

2. 1. 1. 3 Paved taxiway markings

Paved taxiway markings are yellow in color and normally consist of centre line, Taxi-Holding position, Intermediate Taxi-Holding position, edge line and information markings. These markings relay location and other pertinent information and implicit some instructions to pilot.

(a) Centre line marking

Taxiway centre lines are located so as to provide safe clearance between the largest aircraft that the taxiway is designed to accommodate and fixed objects such as buildings, aircraft stands etc. , provided that the pilot of the taxiing aircraft keeps the 'Cockpit' of the aircraft on the centre line and that aircraft on stands are properly parked. Taxi Holding Positions are normally located so as to ensure clearance between an aircraft holding and any aircraft passing in front of the holding aircraft, provided that the holding aircraft is properly positioned behind the holding position. Clearance to the rear of any holding aircraft cannot be guaranteed. When following a taxiway route, pilots need to keep a good lookout, consistent with the prevailing visibility and are responsible for taking all possible measures to avoid collisions with other aircraft and vehicles.

Taxiway Centre line marking consists of a single continuous yellow line marking the declared centre of the taxiway (see Fig 2. 5) . Where a taxiway crosses a runway, the Taxiway Centre line marking indicates the route to be followed but the marking is interrupted as necessary in order to accommodate the runway markings.

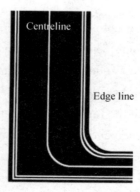

Fig 2. 5   Taxiway centerline and edge line markings

At runway/taxiway intersections, where the taxiway centre line is curved onto the nearside of the runway centre line pilots should take account, where appropriate, of any loss of Runway Declared Distances incurred in following the lead-on line whilst lining up for take-off.

(b) Runway taxi-holding position (RTHP) marking

RTHPs are established on each taxiway leading to a runway in order to protect aircraft on take-off and landing by ensuring that other taxiing aircraft and vehicles are held well clear of the runway and, where appropriate, outside the ILS Sensitive Area. There are two styles of RTHP marking both of which are illustrated in Fig 2.6 and described as follows:

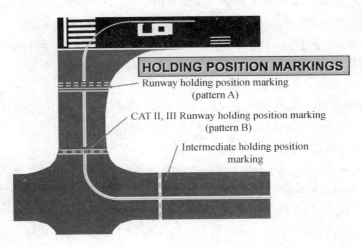

Fig 2.6   Runway taxi-holding position markings

(i) Pattern 'A'

A Pattern 'A' style RTHP marking consists of two solid and two broken lines laid across the entire width of the taxiway and normally at right angles to the taxiway centre line, the broken lines being closer to the runway (see Fig 2.6) .

(ii) Pattern 'B'

A Pattern 'B' style RTHP marking, consists of a ladder style mark laid across the entire width of the taxiway and normally at right angles to the taxiway centre line (see Fig 2.6) .

The last RTHP on a taxiway prior to entering the runway is always marked by a Pattern 'A' RTHP marking; other RTHPs, where established on the same taxiway, are marked by a Pattern 'B' style marking. RTHP markings are also supported by the appropriate RTHP sign.

At those aerodromes where an ATC unit is established, pilots cannot taxi beyond a

Taxi-Holding Position marking towards a runway without ATC clearance.

The direction in which the holding instruction implicit in the Runway taxi-holding position pattern 'B' and intermediate taxi-holding position markings applies, is determined by the accompanying sign, the direction from which the sign face is visible indicates the direction in which the holding requirement applies. The content of those supporting signs will be detailed in the following sections.

(c) Intermediate taxi-holding position (ITHP) marking

At those aerodromes where the taxiway layout is complex or involves multiple intersecting taxiways, ITHPs may be established in order to protect a priority taxiway route. These holding positions are marked by a single broken line laid across the entire width of the taxiway and normally at right angles to the taxiway centre line. An ITHP marking is also supported by a sign. These markings are located so as to provide clearance from aircraft passing in front of the holding aircraft (see Fig 2. 6) .

(d) Taxiway edge marking

Edge markings illustrated at Fig 2. 5, are used where the area beyond the taxiway edge is paved but not normally available for use by aircraft.

(e) Information markings

Information markings, in the form of surface painted directions, may be employed where the use of a sign might cause an unacceptable obstruction.

Upon reaching a Taxi Holding Position identifying a taxi clearance limit, pilot needs to stop the aircraft as close as possible to the Taxi-Hold Position Marking, ensuring that no part of the aircraft protrudes beyond the marking.

2. 1. 1. 4　Paved apron markings

Nowadays there is no international standard for apron markings. Generally, apron markings intended for aircraft maneuvering are yellow in color. Where markings are provided for the guidance of pilots parking aircraft on stands, the position of the marking is determined on the basis that the pilot will endeavor to keep the aircraft nosewheel on the stand centre line. Other colors may be used for markings intended for the guidance of service vehicle drivers.

(a) Stands provided with visual docking guidance

At those aerodromes where visual docking guidance is provided, a variety of different stand layout and markings are used. An example of the layout and markings used at some aerodromes is illustrated in Fig 2. 7.

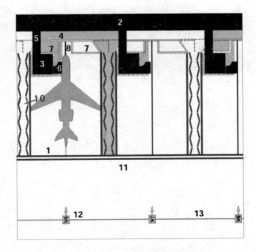

Fig 2. 7   Example of stand layout and markings

1. Stand   2. Pier   3. Gate room   4. Airside road

5. Low bridge   6. Corridor   7. Equipment Parking Area   8. Tug area

9. No parking area   10. Interstand clearway   11. Boundary between apron &taxiway

12. Stand number and centre line   13. Taxi-lane centre line

（b）Self-maneuvering stand markings

The aprons of some aerodromes are provided with surface markings intended to assist pilots in taxiing their aircraft to the correct parking position without the assistance either of a marshaller or of VDGs. These markings are known as Self-maneuvering Stand Markings. A variety of different styles of marking are in use throughout the world. An example of one style is in Fig 2. 8, the method of use is described below:

The pilot turns off the apron taxiway at the arrow which bears the allocated stand number and follows the lead-in line keeping the nosewheel on the centre line. When the FULL TURN arrow is directly abeam the first pilot's position, a turn, using the maximum nose wheel steering angle appropriate to the type of aircraft, is initiated in the direction indicated. This turn is continued until the longitudinal axis of the aircraft is parallel with the alignment bars seen ahead of the aircraft as the designed parking angle is reached. When the aircraft is parallel to the alignment bars, the turn is discontinued and the aircraft permitted to roll forward a distance of not more than 3m in order to straighten the nose wheel, and the aircraft is then stopped. On departure, the pilot taxies off the stand in the direction indicated by the curved lead-off arrow disregarding the alignment bars.

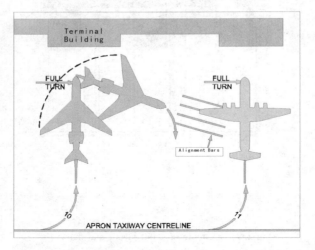

Fig 2. 8   Example of self-maneuvering stand markings

2. 2   Aerodrome signs

The signs located on an aerodrome when used in conjunction with an aerodrome chart are intended to simplify surface movement guidance and control procedures, particularly in conditions of low visibility. Signs are divided into two categories namely Mandatory Signs and Information Signs.

2. 2. 1   Mandatory signs

Mandatory Signs consist of Runway Taxi-Holding Position (RTHP) signs, Intermediate Taxi-Holding Position (ITHP) signs and No Entry signs and display white characters on a red background as illustrated in Fig 2. 9. RTHP and ITHP signs are normally installed alongside surface marking to identify the holding position as well as indicate the direction in which the holding instruction applies. Pilots must not proceed beyond a Mandatory Sign without first obtaining controller's clearance to do so.

Where there is more than one taxiway serving a runway or more than one RTHP on a taxiway, a Location Sign is normally attached to the RTHP sign in order to assist in identifying the position.

(a)

Visual Runway Taxi-Holding Position – denotes the visual Taxi-Holding Position and also the ILS CAT I Holding Position where the Visual and CAT I Holding Positions are co-located.

(i)

(ii)

(b)

CAT I Runway Taxi-Holding Position Sign –denotes ILS CAT I Taxi-Holding Position only where a visual Taxi-Holding Position is established closer to the runway in order to expedite traffic flow.

(i)

(ii)

(c)

CAT II Runway Taxi-Holding Position Sign –marks the ILS CAT II Taxi-Holding Position – a Visual/CAT I Taxi-Holding Position may be established closer to the runway where it is necessary to expedite traffic flow.

(i)

(ii)

(d)

CAT III Runway Taxi-Holding Position Sign –marks the ILS CAT III Taxi-Holding Position – a CAT II Taxi-Holding Position and a Visual/CAT I Taxi-Holding Position may be established closer to the runway where it is necessary to expedite traffic flow.

(i)

(ii)

(e)

Combined Runway Taxi-Holding Position Sign –marks the Taxi-Holding Position where the ILS – Taxi - Holding Positions are co-incident. A Visual or CAT I Taxi-Holding Position Sign may be established closer to the runway where it is necessary to expedite traffic flow.

(i)

(ii)

(f)　**Intermediate Taxi-Holding Position Sign** –marks a Holding Position established to protect a priority route.

(g)　**No entry sign**

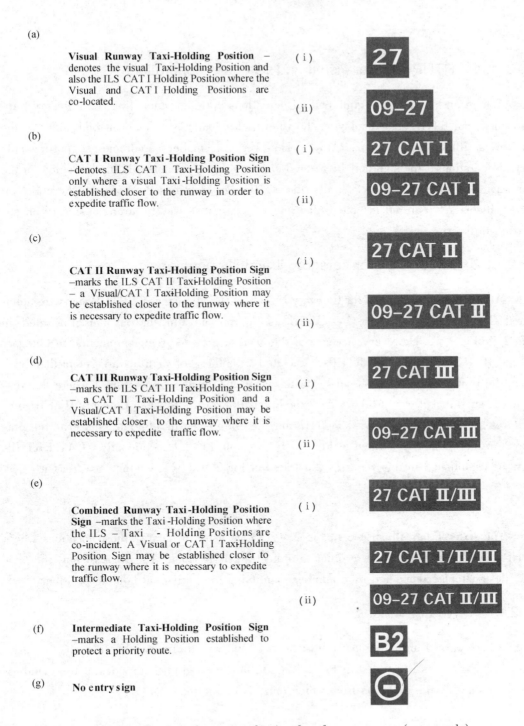

Fig 2. 9　Example of mandatory signs for aircraft surface movements（not to scale）

Note：The signs at（ i ）are used where the taxiway normally serves only one runway direction. The signs at（ii）are used where the taxiway normally serves both runway directions.

2. 2. 1. 1   RTHP sign for visual and category I operations

Where an aerodrome is equipped for operations up to and including ILS Category I approaches, an RTHP sign displaying the runway designator is located on both sides of the taxiway as illustrated in Fig 2. 9 (a) . However, at smaller aerodromes supporting only visual operations, the sign may be located on one side only (normally the left side) of the taxiway. Where there is no ATC unit, the RTHP sign identifies the position where aircraft and vehicles are required to hold whilst conceding right of way to aircraft using or on approach to the runway.

2. 2. 1. 2   RTHP sign for category II and III operations

At aerodromes equipped for Category II and III ILS approaches, RTHP signs are annotated CAT II, CAT III or CAT II/III as appropriate, in the manner illustrated in Fig 2. 9&Fig 2. 10. However, because of the need to provide greater protection to Category II and III ILS systems, the RTHPs associated with these procedures are set farther back from the runway than those associated with visual or Category I operations; where this distance is such that it would hinder the expeditious flow of traffic when Category II or III procedures are not in force, a visual RTHP may be established in addition, closer to the runway, in the manner illustrated at Fig 2. 10 (e) and (f) . Exceptionally, CAT I RTHPs may be established in this way, as illustrated at Fig 2. 9 (b) , for the same reason.

2. 2. 1. 3   ITHP signs

The style of sign illustrated in Fig 2. 9 (f) is used to identify those locations where ITHPs have been established in order to protect a priority route. The signs display the taxiway designator accompanied by a number identifying the individual holding position.

2. 2. 1. 4   No entry signs

Where part of an aerodrome is restricted to one way traffic or is withdrawn from use, No Entry Signs, as illustrated in Fig 2. 9 (g) , are located on both sides of the mouth of the area showing the direction from which entry is prohibited.

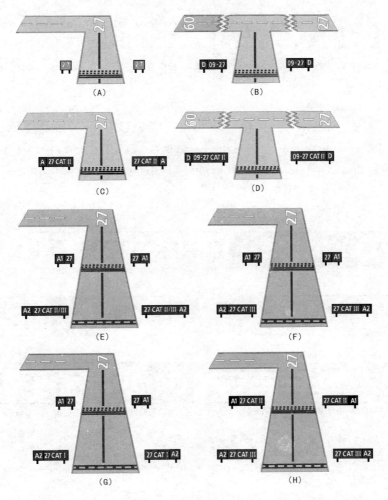

Fig 2. 10　Typical taxi-holding position signs & associated taxiway markings

2. 2. 2　Information signs

Information signs consist of location, direction and destination signs; they are provided only where there is an operational need and should be used in conjunction with an aerodrome chart.

2. 2. 2. 1　Location signs

Location Signs are used to identify taxiways and where necessary (such as at complicated intersections), runways. Taxiways are normally designated by a single letter of the alphabet, e. g. 'A' for taxiway Alpha, 'B' for Bravo etc. The letters 'O', 'I' and 'X' are not used. On larger aerodromes with many taxiways, double letter designators may be used in order to identify minor taxiways adjoining a main route e. g. 'BA' for a minor taxiway adjoining taxiway Bravo. Runway Location Signs use the first two numbers of the runway magnetic heading.

A location sign consists of the characters identifying the runway or taxiway in yellow lettering on a black background surrounded by a yellow border, as illustrated in Fig 2.11. Where there is a need to identify a specific position on a taxiway, a Location Sign, displaying the taxiway designator accompanied by an identifying number as illustrated in Fig 2.11 is used.

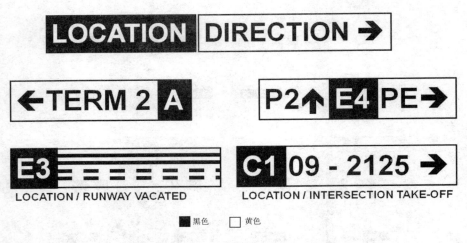

Fig 2.11   Examples of information signs

2.2.2.2   Direction and destination signs

Direction and Destination Signs consist of a route or destination label accompanied by an arrow pointing in the appropriate direction, displayed in black characters on a yellow background as illustrated in Fig 2.11. Direction Signs are normally accompanied by a Location Sign and positioned on the left side of a taxiway or runway before an intersection.

2.2.3   Typical layout of signs

Typical layout patterns of mandatory and information signs are illustrated in Fig 2.10, Fig 2.12 and Fig 2.13.

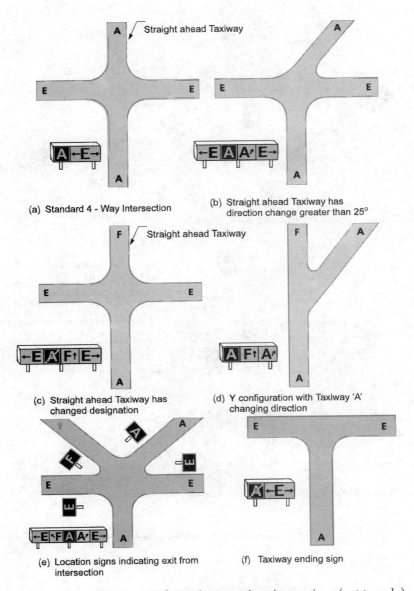

(a) Standard 4 - Way Intersection

(b) Straight ahead Taxiway has direction change greater than 25°

(c) Straight ahead Taxiway has changed designation

(d) Y configuration with Taxiway 'A' changing direction

(e) Location signs indicating exit from intersection

(f) Taxiway ending sign

Fig 2. 12　Examples of taxi guidance signs at taxiway intersections（not to scale）

Note：Signs are laid out as shown above，i. e. from left to right in a clockwise manner. Left turn signs are on the left of the taxiway Location Sign. Right turn signs on the right，except in situations corresponding to（a）&（f）above where the double arrow direction sign is inboard of the taxiway location sign. Adjacent signs are separated by a black vertical delineator.

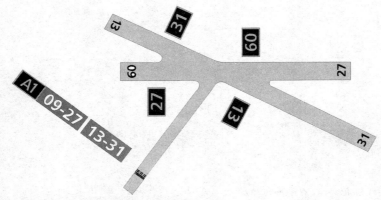

(a) Taxiway Entrance at Intersection of Two Runways

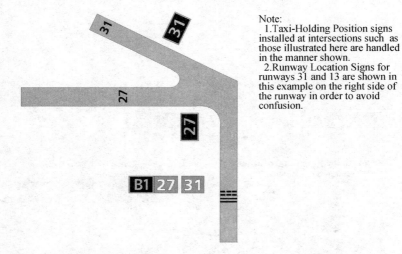

Note:
 1.Taxi-Holding Position signs installed at intersections such as those illustrated here are handled in the manner shown.
 2.Runway Location Signs for runways 31 and 13 are shown in this example on the right side of the runway in order to avoid confusion.

(b) Taxiway Entrance at Intersection of Two Runway

Fig 2. 13   Example of use of runway location signs

Generally, signs are installed together with the surface markings to convey information for pilot; a typical combination has been given in the following figure (see Fig 2. 14).

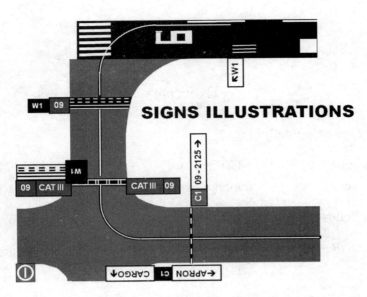

Fig 2. 14　Example of signs installation

2. 2. 4　Aerodrome obstacle marking

Where fixed obstacles are insufficiently conspicuous by day, they are normally marked either by alternating bands or by a checked pattern of red or orange and white. Vehicles and other mobile equipment frequently employed on the movement area are normally painted yellow or alternatively may be fitted with distinctive yellow markers or flags. Unserviceable parts of the movement area are normally delineated either by marker board painted in alternate bands of red or orange and white, or by diagonally split orange/white flags. Where practicable, an unserviceable part of the movement area would also be marked by one or more large white crosses. Methods of marking obstacles are shown in Fig 2. 15.

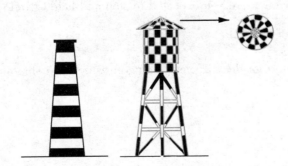

Fig 2. 15　Example of marking on buildings and structures

Exercise

I. Explain the following definitions.

(1) mandatory signs   (2) RTHP marking   (3) ITHP Marking
(4) touchdown zone   (5) runway designation marking   (6) aiming point

II. Answer the following questions.

(1) According to the purposes of the surface markings, what are the two groups of surface markings?

(2) What's the difference between the holding position markings pattern A and B? Please draw out their pattern.

(3) What's displaced threshold? Please draw out displaced threshold marking.

III. Suppose you were a pilot on taxiway L (look at the following figure), you are given an instruction from aerodrome controller "Taxi Runway 18 via Lima, hold short Runway 18 ILS critical area." Answer the following questions.

(1) At what point on Lima are you expected to stop and hold? Circle it out on the figure.

(2) Draw out the pattern of the surface marking on which you stop and hold.

Chapter 2 Aerodrome surface markings and signs

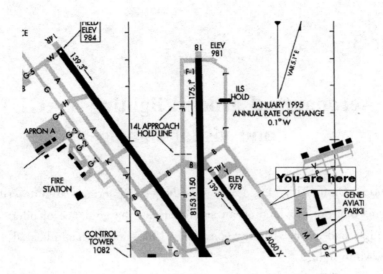

Ⅳ. **Suppose you were a pilot exiting runway 22R, at which hold line you have to stop? Circle out that line on the following figure.**

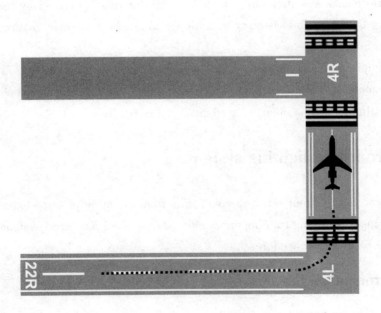

Chapter 3

Aeronautical ground lighting (AGL) and visual signals

Aeronautical Ground Lighting (AGL) is the generic term used to describe the various lighting systems that are provided on an aerodrome for the guidance of pilots operating aircraft both at night and in low visibility conditions. Unless otherwise indicated, AGL systems emit a steady white light.

High intensity AGL systems that are provided in support of low visibility operations normally have facility for the independent control of brilliancy of each element of the system. The intensities are set up by aerodrome controller in order to suit local conditions and a pilot may ask controller to adjust them if they are found to be inappropriate.

The performance specification of high intensity lighting is defined by the need to provide guidance by day in low visibility conditions; the highest intensity settings are normally used in those conditions.

Lower intensities are normally used by night. Low intensity systems are provided at those aerodromes at which operations are conducted at night but not in low visibility conditions; the brilliancy of low intensity systems is not normally adjustable.

3. 1　Aerodrome lighting system

Aerodrome light system varies in complexity from the basic patterns found at small aerodromes in support of flying training operations, to the more advanced systems used in support of Category Ⅲ ILS procedures.

3. 1. 1　Aeronautical beacons

An Aeronautical beacon are normally provided at those aerodromes that operate at night and where the level of background lighting, the surrounding terrain, the proximity of other aerodromes or the lack of navigation aids make the aerodrome difficult to locate or to identify. There are two types of Aeronautical beacons, Aerodrome Beacon and Identification Beacon. An identification beacon replaces an aerodrome beacon at an aerodrome which is intended for use at night and cannot be easily identified from nearby aerodromes in the air by other means.

3. 1. 1. 1 Aerodrome beacon

An Aerodrome Beacon is normally provided at an aerodrome that is situated well away from other aerodromes and where no confusion exists as to identity. It shows either colored flashes alternating with white flashes, or white flashes only. Where used, the colored flashes emitted by beacons at land aerodromes are green and yellow at water aerodromes.

3. 1. 1. 2 Identification beacon

An identification beacon flashing the aerodrome identification code in the International Morse Code is normally provided at an aerodrome where a number of aerodromes in the same vicinity operate at night and confusion could arise as to identity. It shows flashing green at a land aerodrome and flashing-yellow at a water aerodrome.

3. 1. 2 Approach lighting system

A variety of approach lighting system based on the centre line and cross bar, is in use at aerodromes throughout the world. These systems range from the simple low intensity centre line and cross bar intended to serve visual runways at night only, to the more complex Calvert System comprising centre line and 5 cross bars (CL5B) required for day and night use on ILS equipped runways. At some aerodromes where, because of high levels of background lighting, the approach lighting is difficult to pick out at distance, strobe lighting is provided either in addition to the standard approach lighting or on its own. Except where supplemented by red side barrettes, approach lighting is white in color.

3. 1. 2. 1 Simple approach lighting system

A simple approach lighting system includes a row of lights on the extended centre line of the runway extending, whenever possible, over a distance of not less than 420 m from the threshold with a row of lights forming a crossbar 18 m or 30 m in length at a distance of 300 m from the threshold (See Fig 3. 1).

Simple approach lighting system type A with red color intends to serve non-instrument runways and type B in white color is installed to serve non-precision runways at nights.

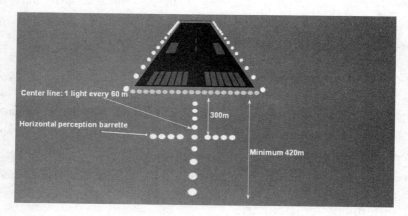

Fig 3. 1   Simple approach light system

3. 1. 2. 2   Precision approach category I lighting system

A precision approach category I lighting system includes a row of lights on the extended centre line of the runway extending, wherever possible, over a distance of 900 m from the runway threshold with a row of lights forming a crossbar 30 m in length at a distance of 300 m from the runway threshold. The centre line and crossbar lights of the lighting system are fixed lights showing variable white. Each centre line light position consists of several barrettes (see Fig 3. 2) .

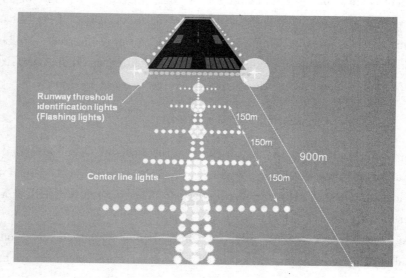

Fig 3. 2   Precision approach lightings system CAT I

3. 1. 2. 3   Approach lighting system for Cat II and III ILS approaches

At those aerodromes where Category II and III ILS approaches are conducted, Approach Lighting consisting of white centre line barrettes and two rows of red side barrettes, (an example is shown in Fig 3. 3), is installed in order to provide the pilot with enhanced visual cues over the last 300m of the approach.

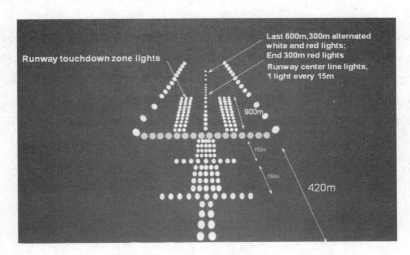

Fig 3. 3   Precision approach and runway lighting system CAT II

3. 1. 2. 4   PAPI (APAPI) and T-VASI

Precision Approach Path Indicator (PAPI) provides visual approach slope guidance by use of red and white light signals. The system normally comprises a single row of 4 light units except that on those runways without ILS, and a 2 unit system (APAPI) is used. It's normally installed on the left side of the runway as seen from the approach. However, at some aerodromes, additional units are added on the right side in order to give better roll guidance. The signals are interpreted as illustrated in Fig 3. 4.

Used together with ILS, PAPI is to ensure, as far as practicable, correlation between the two glide paths. However, such a siting is made on the assumption that the pilot's eye level is above the ILS glide path receiver aerial, as is the case with most commercial aircraft. Pilots of aircraft in which the ILS aerial is mounted above the level of the pilot's eye (e. g. Shorts 330) may see a PAPI indication 'slightly low' when on the ILS glide path.

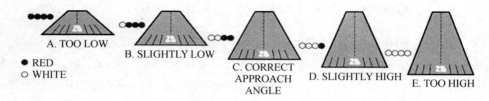

Fig 3. 4   View of PAPI from approach

Although PAPI is now in widespread use throughout the ICAO States, there are many countries, where Visual Approach Slope Indicator (VASI) is used. Both PAPI and VASI are ICAO approved equipment. The VASI operates on the same principle as the PAPI in that it provides a pilot with guidance in relation to the correct approach slope. The 2 Bar VASIs provide approach slope guidance and the interpretation of the signal is illustrated in Fig 3. 5.

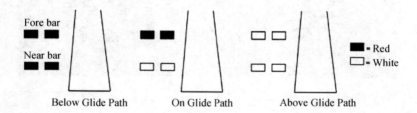

Fig 3. 5   View of 2 bar VASI from approach

3. 1. 3   Runway lightings

All runways licensed for night use have Edge, Threshold and End Lighting. Centre line and Touchdown Zone Lighting is provided as additional guidance in support of low visibility operations.

3. 1. 3. 1   Runway edge lighting

Runway Edge Lighting is normally located along the edges of the area declared for use as the runway. However, where a paved surface is wider than the declared runway width, the lights are located at the edge of the pavement and the declared width delineated by white edge markings. Edge lighting may be provided either by elevated or by flush fitting lamp fixtures. At some aerodromes where elevated runway edge lights are employed, the light fixtures are located on the grass shoulder beyond the declared runway width (see Fig 3. 6).

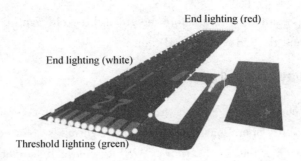

End lighting (red)

End lighting (white)

Threshold lighting (green)

Fig 3. 6   Runway edge lightings

Runway Edge Lighting is showing white except in the following instances：

（a）Caution zone lighting

On ILS equipped runways without centre line lighting, yellow edge lighting is installed on the upwind 600m or one third of the lighted runway length available, whichever is the less. The yellow 'caution zone' so formed gives a visual warning of the approaching runway end.

（b）Pre-threshold lighting

Where a landing threshold is displaced, but the pre-threshold area is available for the take-off run, the lights between the beginning of the runway pavement and the displaced threshold show red from the approach. Pilots taking off in such a situation see red edge lights up to the green threshold then white edge lights beyond. Where a starter extension, narrower than its associated runway is provided, blue edge lighting is normally used to mark the edges.

（c）Runway exit lighting

One or two omni-directional blue lights may replace or supplement the edge lights in order to indicate an exit taxiway.

（d）Stopway lighting

Where stopway is provided at the end of a runway, the declared stopway is delineated by red edge and end lighting showing only in the direction of landing. A stopway is provided for emergency use only and is not normally suitable for routine use.

3. 1. 3. 2   Runway threshold and runway end lighting

Runway threshold lighting is green and indicates the start of the available landing distance. Green threshold wing-bars are provided at certain aerodromes where there is a need to accentuate the threshold. Runway end lighting is red and marks the extremity of the runway that is available for maneuvering (see Fig 3.6).

Pilots are not permitted to land before the green threshold lighting nor continue a landing roll or taxi beyond the red runway end lights.

3. 1. 3. 3　Runway centre line lighting

High intensity centre line lighting is provided in addition to edge lighting on runways equipped for low visibility operations. The centre line lighting is color coded in order to warn a pilot of the approaching end of the runway; White centre line lighting extends from the threshold to 900m from the runway end, the following 600m is lit with alternate white and red lights, and the final 300m lit by red centre line lighting, as shown in Fig 3.3.

3. 1. 3. 4　Touchdown zone lighting

On runways equipped for Category II and III approaches, additional lighting consisting of two rows of white barrettes, as shown in Fig 3.3 is installed in order to provide textural cues in the touchdown area. The additional lighting extends from the threshold either for 900m or to the midpoint of the runway whichever is the lesser distance.

The length of the Touchdown Zone lighting (normally 900m) determines the length of the Obstacle Free Zone established to protect CAT II & III approaches below DH and in the event of a baulked landing ('go around') after DH. A 'go around' initiated beyond the end of the Touchdown Zone lighting is unlikely to be contained within the Obstacle Free Zone.

3. 1. 4　Taxiway lighting

3. 1. 4. 1　Taxiway centre line lights

At those aerodromes equipped for low visibility operations, taxiway centre line lights showing steady green are provided on an exit taxiway, taxiway, de-icing and anti-icing facility and apron intended for use in runway visual range less than 350m, as shown in Fig 3.7. Green taxiway centre line lighting sometimes is also provided on the runway prior to an exit taxiway in order to give lead-off guidance.

Where centre line lighting is installed on a taxiway leading onto a runway, the taxiway lighting is curved onto the near side of the runway centre line and pilots should make an appropriate allowance for any loss of Runway Declared Distance incurred in following the 'lead-on' lighting whilst lining up for take-off.

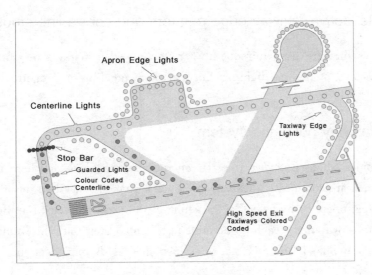

Fig 3. 7   Taxiway lighting

Where part of a taxiway equipped with centre line lighting lies within the ILS Sensitive Area or is sufficiently close to a runway that aircraft on that part of the taxiway would present an obstruction to aircraft landing or taking-off, that part of the taxiway will be identified by alternate green and yellow centre line lights. Pilots need to avoid stopping with any part of their aircraft in such areas.

3. 1. 4. 2   Taxiway edge lights

Taxiway edge lights with steady blue are provided at the edges of a holding bay, de-icing and anti-icing facility, apron, etc. intended for use at night and on a taxiway not provided with taxiway centre line lights and intended for use at night, even where green centre line lighting is provided, blue taxiway edge lighting may also be installed as additional guidance on sections of taxiway that are difficult to negotiate (see Fig 3. 7) .

The edges of aprons, turning and holding areas are normally marked by blue lighting.

3. 1. 4. 3   Taxiway intersection lights

At some aerodromes where multiple intersecting taxiways are not provided with selective route guidance, Taxiway Intersection Lights are provided. These lights consist of a row of at least 3 steady yellow lights disposed symmetrically about the taxiway centre line. Pilots approaching an intersection where these lights are displayed gives way to crossing traffic unless otherwise instructed by ATC.

3. 1. 4. 4   Reflective taxiway edge markers and centre line studs

On taxiways that are used infrequently, reflective edge markers or centre line studs may be used instead of taxiway lighting. Edge markers are blue and centre line studs are green.

3. 1. 4. 5   Taxiway guidance system

At aerodromes where Category II & III operations take place or where ground movement requirements are complex, a taxiway guidance system are installed in order to regulate traffic. The system operates by selective switching of the taxiway centre line lighting so that individual sections or routes, each terminating at a lighted Stop Bar, are illuminated in order to show the way ahead. The Stop Bar is extinguished as the next section of taxiway centre line lighting is selected.

3. 1. 5   Stop bars and Lead-on lights

Lighted Stop Bars and Lead-on Lights are provided at those aerodromes authorized for low visibility operations (RVR less than 550m). A Stop Bar consists of a row of lights spaced equally across the taxiway normally at right angles to the centre line and showing red towards an approaching aircraft (see Fig 3. 8). At some aerodromes where, for example, a Stop Bar is located on or close to a bend in the taxiway route, additional elevated red lights are installed outboard of each taxiway edge, in order to provide maximum advanced warning of the Stop Bar location. Stop Bars are normally installed in association with green Lead-on Lights which form part of the taxiway centre line lighting beyond the stop bar. The Lead-on Lights are interlinked with the Stop Bar so that when the Stop Bar is 'on' the green centre line beyond the Stop Bar is 'off' and vice versa. In this way, the Stop Bar and associated Lead-on Lights act in the same sense as traffic lights and thus pilots needn't taxi an aircraft across a lighted Stop Bar.

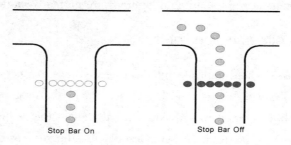

Fig 3. 8   Stop bar and leading on lights

3. 1. 6　Intermediate holding position lights

Except where a stop bar has been installed, intermediate holding position lights are provided at an intermediate holding position intended for use in runway visual range conditions less than a value of 350m. Intermediate holding position lights includes three fixed unidirectional lights showing yellow in the direction of approach to the intermediate holding position with a light distribution similar to taxiway centre line lights if provided. The lights will be disposed symmetrically about and at right angle to the taxiway centre line.

3. 1. 7　Runway guard lights

On aerodromes equipped for low visibility operations, all runway entry points are "protected" by Runway Guard Lights. These are pairs of alternately flashing yellow lights; one pair located on each side of the taxiway and provides a warning of the close proximity of the runway. Where the taxiway is wider than normal, an alternative form of Runway Guard Light may be provided comprising additional pairs of flashing yellow lights inset into and stretching across the full width of the taxiway. The electrical circuits are so arranged that alternate lights flash in unison. Runway Guard Lights, often referred to as "Wig Wags", are illustrated in Fig 3. 9.

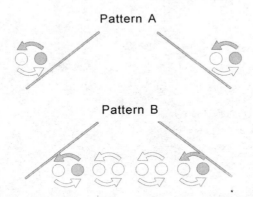

Fig 3. 9　Taxiway lighting & runway guard Lights

3. 1. 8　Aerodrome obstacles lighting

All objects located above a take-off climb surface, approach surface, transitional surface within 3 000 m of the inner edge of the surface, and those above horizontal surface or adjacent to a take-off climb surface, are marked and lighted with steady if the runway is used at night.

Fixed obstacles of 45m or less in height, width and length are normally lit by a single steady red light placed at the highest practicable point; those obstacles of greater size are normally provided with additional red lights in order to outline the extent of the obstruction as shown in Fig 3. 10 (a) and (b). Surface obstructions and unserviceable parts of the movement area are normally delineated by portable red lights. Mobile obstacles such as vehicles and equipment frequently employed on the movement area normally display a yellow flashing light except that emergency service vehicles responding to an incident display flashing blue lights.

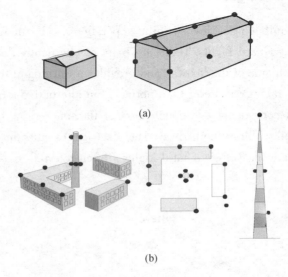

(a)

(b)

Fig 3. 10   Obstacle lighting

3. 2   Visual signals

3. 2. 1   Visual ground signals

A Signals area is employed in order to provide information relating to the conduct of flying operations. Where provided, the Signals Area, measuring approximately 12m square and bounded by a white border, is so located on the aerodrome that it is visible from all directions of approach (see Fig 3. 11) .

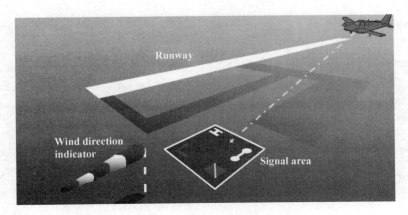

Fig 3. 11   Signal area of an aerodrome

Meanings of visual signals displayed on the Aerodrome are described in Fig 3. 12.

Signal/Marking	Location	Description & Meaning
	On the specified Signal aera.	A white landing T signifies that aircraft and gliders taking-off or landing shall do so in a direction parallel with the shaft of the T and towards the cross arm, unless otherwise authorized by the appropriate ATC unit.
	On the specified Signal area.	A white disc displayed alongside the cross arm of the T in line with the shaft of the T signifies that the direction of landing and take-off do not necessarily coincide.
	On the specified Signal area.	A red and yellow striped arrow placed along the whole of two adjacent sides of the signals area and pointing in a clockwise direction signifies that a right hand circuit is in force.
	On the specified Signal area.	A red panel square with a yellow diagonal stripe signifies that owing to the bad state of the maneuvering area, or for any other reason, special precautions must be observed in approaching to land or in landing.

Signal/Marking	Location	Description & Meaning
	On the specified Signal area.	A red panel square with yellow stripes along each diagonal signifies that the aerodrome is unsafe for the movement of aircraft and that landing is prohibited.
	On the specified Signal aera.	A horizontal white dumb-bell when displayed in a signal area indicates that aircraft are required to land, take off and taxi on runways and taxiways only.
	On the specified Signal area.	A black strip across each disc of the white dumb-bell at right angles to its shaft indicates that aircraft are required to land and take off on runways only, but other manoeuvres need not be confined to runways and taxiways.
	On the specified Signal area.	A red letter L displayed on the dumb-bell signifies that light aircraft are permitted to take-off and land either on a runway or on the area on the aerodrome designated by a large white letter L.
	On the specified Signal area.	A double white cross displayed horizontally in the signal area indicates that the aerodrome is being used by gliders and that glider flights are being performed.
	On the landing area.	A white letter H signifies that helicopters shall take-off and land within the area designated by a large white letter H.
	Aerodrome administrative building/ control tower/ flight plan office.	The letter C displayed vertically in black against a yellow background indicates the location of the air traffic services reporting office.

Signal/Marking	Location	Description & Meaning
	At intervals along the boundary of an aerodrome.	Orange and white striped markers are used to delineate the boundary where it is insufficiently conspicuous.
	Normally located on the control tower building.	A panel or flag consisting of red and yellow squares indicates that aircraft may taxi only in accordance with ATC instructions.
	On the landing area.	A large white letter L indicates an area normally of grass reserved for landing and take-off of light aircraft.
	On a selected part of the aerodrome.	A yellow marker in the shape of a Cross indicates an area reserved for the dropping of tow ropes or similar articles.
	On a portion of a runway.	A white mark in the shape of an elongated Cross indicates that that portion of the runway up to the next standard marking is unfit for use by aircraft.
	On a portion of a taxiway.	A yellow mark in the shape of a Cross indicates that that portion of the taxiway up to the next standard marking is unfit for use by aircraft.
	Normally located on the control tower building.	A black two figure designator against a yellow background indicates the runway in use / direction of takeoff and landing.
	Adjacent to the landing area and visible from each runway threshold and all directions of approach. Sometimes lit at night.	Orange (or other conspicuous colour) wind sleeve indicates wind direction and speed.

Fig 3. 12 Meanings of visual signals

3.2.2 Light and flare signals

Under some circumstances, aerodromes, especially at those where general aviation movements are significant, light signals and pyrotechnic signals are used to convey instructions to pilots and ground staff and have the meanings described in Fig 3.13.

Signal		Meaning
Signal Red flare from tower. Steady red light to aircraft or vehicle as indicated. Red flare from aircraft.		**Meaning** -Notwithstanding any previous instructions, do not land for the time being. -Give way to other aircraft and continue circling. Stop(for aircraft or vehicles on ground). -Immediate assistance required.
Signal Flashing red light to aircraft or vehicle.		**Meaning** - Aerodrome unsafe; Do not land. -Move clear of landing area in use.
Signal Flashing green light to aircraft or vehicle.		**Meaning** -return to aerodrome await landing clearance. -Cleared to taxi.
Signal Steady green light to aircraft.		**Meaning** -Cleared to land. -Cleared to take-off.
Signal Steady or flashing green or green flare from aircraft.		**Meaning** -By Night – may I land. -By Day – may I land in a direction different from that indicated.

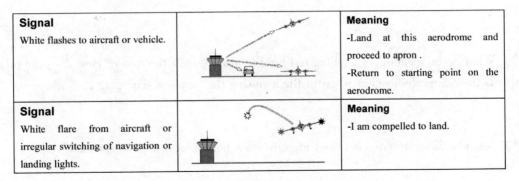

Signal		Meaning
White flashes to aircraft or vehicle.		-Land at this aerodrome and proceed to apron . -Return to starting point on the aerodrome.
Signal White flare from aircraft or irregular switching of navigation or landing lights.		**Meaning** -I am compelled to land.

Fig 3. 13　Light and flare signals

The acknowledgement by an aircraft to the abovementioned signals are used as followings:

(a) When in flight :

During the hours of daylight:

— by rocking the aircraft's wings;

During the hours of darkness:

— by flashing on and off twice the aircraft's landing lights or, if not so equipped, by switching on and off twice its navigation lights.

(b) When on the ground:

During the hours of daylight:

— by moving the aircraft's ailerons or rudder;

During the hours of darkness:

— by flashing on and off twice the aircraft's landing lights or, if not so equipped, by switching on and off twice its navigation lights.

Exercise

I. Explain the following definitions.

(1) aerodrome beacon　　(2) PAPI　　(3) taxiway guidance system

(4) stopbar lights　　(5) VASI　　(6) simple approach lighting system

II. Answer the following questions.

(1) For an advanced complex aerodrome, Aeronautical Ground Lighting (AGL) generally consists of various lighting systems, what are they and what's the main function of them?

(2) What's the meaning of flashing red light to aircraft from the tower? How does the pilot express the acknowledgement when in flight during the hours of daylight?

(3) What's the difference between identification beacon and location beacon?

(4) What's the color and installing position for the stopway lighting?

III. Mark out the color code of the lights on the following figure.

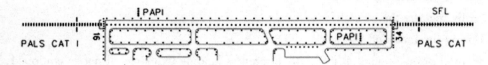

PAPI

SFL

PALS CAT I

PAPI

PALS CAT

(1) Aerodrome lighting system

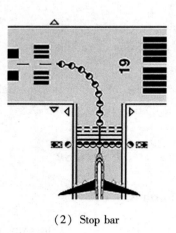

19

(2) Stop bar

Chapter 4

Essential aerodrome information

4. 1　Essential information on aerodrome conditions

Essential information on aerodrome conditions is necessary to safety in the operation of aircraft, which pertains to the movement area or any facilities usually associated there with. For example, construction work on a taxi strip not connected to the runway-in-use is not essential information to any aircraft except one that might be taxied in the vicinity of the construction work. Another example, if all traffic is confined to runways, it is considered as essential aerodrome information to any aircraft not familiar with the aerodrome.

Essential information on aerodrome conditions generally includes information relating to the following:

(a) construction or maintenance work on, or immediately adjacent to the movement area;

(b) rough or broken surfaces on a runway, a taxiway or an apron, whether marked or not;

(c) snow, slush or ice on a runway, a taxiway or an apron;

(d) water on a runway, a taxiway or an apron;

(e) snow banks or drifts adjacent to a runway, a taxiway or an apron;

(f) other temporary hazards, including parked aircraft and birds on the ground or in the air;

(g) failure or irregular operation of part or all of the aerodrome lighting system;

(h) any other pertinent information.

Up-to-date information on the conditions on aprons is always available to the aerodrome controller. The responsibility of the aerodrome control tower in relation to aprons is limited to the transmission to aircraft of the information which is provided to it by the authority responsible for the aprons.

It's aerodrome controller's responsibility to transmit essential information on aerodrome conditions to all aircraft, except when it is known that the aircraft already has received all or part of the information from other sources. Controller ensures the information is given in sufficient time for pilot to make proper use of it, and the hazards are identified as distinctly as possible.

When a not previously notified condition pertaining to the safe use by aircraft of the maneuvering area is reported to or observed by the controller, he/she has the responsibility to inform the aerodrome authority until operations on that part of the maneuvering area terminated or advised by aerodrome authority.

4.1.1 Extracting aerodrome condition information

Whenever information is provided on aerodrome conditions, it is done in a clear and concise manner so as to facilitate appreciation by pilot of the situation described. Controller issues it whenever deemed necessary in the interest of safety, or when requested by pilot. If controller initiates the information, he/she transmit to each aircraft concerned in sufficient time to enable the pilot to make proper use of the information.

The following terms are used by aerodrome controller transmitting to pilot concerned information that water is present on a runway:

DAMP — the surface shows a change of color due to moisture.

WET — the surface is soaked but there is no standing water.

WATER PATCHES — patches of standing water are visible.

FLOODED — extensive standing water is visible.

4.2 Aerodrome meteorological information

Prior to taxiing for take-off, the controller advise aircraft the following elements of weather information, in the order listed, with the exception of such elements which it is known the aircraft has already received.

（a）the runway to be used;

（b）the surface wind direction and speed, including significant variations;

（c）the QNH altimeter setting and, either on a regular basis in accordance with local arrangements or if so requested by the aircraft, the QFE altimeter setting;

（d）the air temperature for the runway to be used, in the case of turbine-engined aircraft;

（e）the visibility representative of the direction of take-off and initial climb, if less than 10 km, or, when applicable, the RVR value（s）for the runway to be used.

Prior to take-off, controller advises pilot the followings:

(a) any significant changes in the surface wind direction and speed, the air tempera-ture, and the visibility or RVR value (s);

(b) significant meteorological conditions in the take-off and climb-out area, except when it is known that the information has already been received by pilot. Significant meteor-ological conditions here include the occurrence or expected occurrence of cumulonimbus or thunderstorm, moderate or severe turbulence, wind shear, hail, moderate or severe ic-ing, severe squall line, freezing precipitation, severe mountain waves, sand storm, dust storm, blowing snow, tornado or waterspout in the take-off and climb-out area.

Prior to entering the traffic circuit or commencing its approach to land, controller pro-vides pilot the following elements of information, in the order listed, with the exception of such elements which it is known the pilot has already received:

(a) the runway to be used;

(b) the surface wind direction and speed, including significant variations therefrom;

(c) the QNH altimeter setting and, either on a regular basis in accordance with local arrangements or, if so requested by pilot, the QFE altimeter setting.

4.2.1　Extracting meteorological information

The ATS unit extracts concerned meteorological information concerning the meteoro-logical conditions from the following meteorological messages, which is provided by the me-teorological office and transmits to aircraft in accordance with relevant regulations:

(a) local meteorological routine and special reports;

(b) meteorological reports in the METAR/SPECI code forms, for dissemination to other aerodromes beyond the aerodrome of origin (mainly intended for flight planning, VOLMET broadcasts and D-VOLMET).

4.2.1.1　Mean surface wind direction, speed and significant variations

Information on surface wind direction is referenced to degrees true north. Information on surface wind direction obtained from surface wind indicator and passed to pilots by con-trollers is given in degrees magnetic.

In meteorological reports, the direction is given in true degrees and the speed in km/ h (kt). All directional and speed variations refer to the preceding 10-minute peri-od. Directional variation is given when the total variation is 60 degrees or more; when the mean speed is above 6 km/h (3 kt) and the wind varies less than 180 degrees, it is ex-pressed as the two extreme directions between which the wind has varied; otherwise, it is indicated as VRB, followed by the mean speed, with no indication of the mean wind direction. Speed variations (gusts) are reported only when the variation from the mean speed is 20 km/h (10 kt) or more.

In local meteorological routine and special reports:

(a) the averaging period for mean surface wind direction and speed are 2 minutes;

(b) speed variations are expressed as the maximum and minimum values attained;

(c) light and variable surface winds of 6 km/h (3kt) or less include a range of wind directions, whenever possible.

In meteorological reports disseminated beyond the aerodrome:

(a) the averaging period for mean surface wind direction and speed are 10 minutes;

(b) speed variations are expressed as the maximum value attained. Minimum wind speed are not be included.

4. 2. 1. 2 Visibility

In local meteorological routine and special reports, the visibility is representative of:

(a) the take-off/climb-out area for departing aircraft;

(b) the approach and landing area for arriving aircraft.

In meteorological reports disseminated beyond the aerodrome, the visibility is representative of the aerodrome and its immediate vicinity. In case of significant directional variations in visibility:

(a) the lowest visibility need to be reported;

(b) additional values need to be given with indications of the direction of observation.

When the visibility is less than 500 m, it then is expressed in steps of 50 m; when it is 500 m or more but less than 5000 m, in steps of 100 m; 5000 m or more but less than 10 km, in kilometer steps; and when it is 10km or more, only 10 km shall be given, except when the conditions for the use of CAVOK apply.

4. 2. 1. 3 Runway visual range (RVR)

Runway visual range values up to 400 m are given in increments of 25 m, values between 400 and 800 m in increments of 50 m, and values above 800 m are given in increments of 100 m. Runway visual range values which do not fit the reporting scale will be rounded down to the nearest lower step in the reporting scale.

In local meteorological routine and special reports, the averaging period for RVR is 1 minute and:

(a) when the runway visual range is above the maximum value which can be determined by the system in use, it need to be reported as more than the specified distance, e. g. RVR RWY 14 ABV 1 200M where the figure 1 200 is the maximum value that can be determined by the system; or when the runway visual range is below the minimum value which can be measured with the system in use, it need to be reported as less than the specified distance, e. g. RVR RWY 10 BLW 150M.

(b) when the runway visual range is observed from one location along the runway about 300 m from the threshold, it need to be included without any indication of location, e. g. RVR RWY 20 600M; or when the runway visual range is observed from more than one location along the runway, the value for the touchdown zone need to be given first and then be followed by the values representative of the mid-point and stop-end. The locations for which these values are representative need to be given in meteorological reports as TDZ, MID and END, e. g. RVR RWY 16 TDZ 600M MID 400M END 400M; Where reports for three locations are given, the indication of these locations may be omitted, provided that the reports are passed in the order specified above, e. g. RVR RWY 16 600M 400M 400M.

(c) when there is more than one runway in use, the available runway visual range values for each runway is given and the runways to which the values refer need to be indicated, e. g. RVR RWY 26 800M; RVR RWY 20 700M; if the runway visual range is available only for one runway, that runway need to be indicated, e. g. RVR RWY 20 600M.

In meteorological reports disseminated beyond the aerodrome, the averaging period for RVR is about 10 minutes and:

(a) only the value representative of the touchdown zone need to be given, and no indication of location on the runway is included;

(b) when there is more than one runway available for landing, touchdown zone runway visual range values need to be included for all such runways, up to a maximum of four, and the runways to which the values refer is indicated, e. g. RVR RWY 26 500M RVR RWY 20 800M;

(c) when the runway visual range values during the 10-minute period immediately preceding the observation have shown a distinct tendency, such that the mean during the first 5 minutes varies by 100 m or more from the mean during the second 5 minutes of the period, this is indicated by the abbreviation "U" for an upward tendency and by the abbreviation "D" for the downward tendency, e. g. RVR RWY 12 300M/D;

(d) when the fluctuations of the runway visual range during the 10-minute period immediately preceding the observation have shown no distinct tendency, it will be indicated by the abbreviation "N";

(e) when the one-minute runway visual range values during the 10-minute period immediately preceding the observation vary from the mean value by more than 50 m or more than 20 per cent of the mean value, whichever is the greater, the one-minute mean minimum and the one-minute mean maximum values are included instead of the 10-minute mean value, e. g. RVR RWY 18 MNM700M MAX1100M.

4. 2. 1. 4   Present weather

The types of the present weather phenomena is given in meteorological reports in terms of drizzle, rain, snow, snow grains, ice pellets, ice crystals (diamond dust), hail, small hail and/or snow pellets, fog, mist, sand, dust (widespread), haze, smoke, volcanic ash, dust/sand whirls (dust devils), squall, funnel cloud (tornado or waterspout), dust storm and sand storm.

The following characteristics of the present weather phenomena need to be included, as appropriate, in connection with the types listed in abovementioned: thunderstorm, shower, freezing, blowing, low drifting, shallow, patches and partial.

The relevant intensity (light, moderate, heavy) or, as appropriate, the proximity to the aerodrome (vicinity) of the reported present weather phenomena will be used.

4. 2. 1. 5   Amount and height of base of low cloud

Cloud amount is given in order that FEW (1-2 oktas), SCT (3-4 oktas), BKN (5-7 oktas) or OVC (8 oktas), type [only if cumulonimbus (CB) or towering cumulus (TCU)] and height of base in metres (feet) . If the base of the lowest cloud is diffuse or ragged or fluctuating rapidly, the minimum height of the cloud or cloud fragments are given together with an appropriate description of the characteristics thereof. If there are no clouds and no restriction on vertical visibility, and the abbreviation CAVOK is not appropriate, SKC can be used. If there are no clouds below 1500 m (5 000 ft) or below the highest minimum sector altitude, whichever is greater, no cumulonimbus and no restriction on vertical visibility and the abbreviations "CAVOK" and "SKC" are not appropriate, the abbreviation "NSC" can be used. When the sky is obscured, the vertical visibility shall be given, if available.

4. 2. 1. 6   Air temperature and dew point

Air temperature and dew point temperature are given rounded to the nearest whole degree Celsius, with observed values involving 0. 5℃ rounded to the next higher whole degree Celsius.

4. 2. 1. 7　Altimeter setting（s）

Generally, aerodrome controller issues the QNH altimeter setting to pilot. The QFE altimeter setting is also be available and passed either on a regular basis in accordance with local arrangements or if requested by the pilot. Altimeter settings is given in hectopascals in four digits together with the unit of measurement used and is rounded down to the nearest lower whole hectopascal.

4. 2. 1. 8　Significant meteorological information

Significant meteorological information includes any available information on meteorological conditions in the area of the aerodrome, approach, missed approach or climb-out areas relating to the location of cumulonimbus or thunderstorm, moderate or severe turbulence, wind shear, hail, severe squall line, moderate or severe icing, freezing precipitation, severe mountain waves, sand storm, dust storm, blowing snow, tornado or waterspout, as well as any information on recent weather of operational significance (i. e. freezing precipitation; moderate or heavy precipitation; moderate or heavy blowing snow; duststorm or sandstorm; thunderstorm; tornado or waterspout; volcanic ash) observed during the period since the last issued routine report or last hour, whichever is shorter, but not at the time of observation.

In meteorological reports disseminated beyond the aerodrome, only information on wind shear and on recent weather of operational significance as given in abovementioned is included.

When the visibility is 10km or more, there is no cloud below 1500 m (5000 ft), or below the highest minimum sector altitude, whichever is greater, and no cumulonimbus, and there is no weather of significance types of the present weather phenomena, information on visibility, runway visual range, present weather and cloud amount, type and height is replaced by the term "CAVOK".

4. 3　Other information

4. 3. 1　Start-up Time

When so requested by the pilot prior to engine start, aerodrome controller issues an expected take-off time in accordance with its allocated slot time. When controller anticipate delay for a departing aircraft is less than a certain time period, he/she then clears the aircraft to start-up at its own discretion, and when delay for a departing aircraft is anticipated to exceed a time period, controller issues an expected start-up time to pilot requesting start-up.

4.3.2 Essential local traffic information

Essential local traffic includes any aircraft, vehicle or personnel on or near the manoeuvring area or traffic operating in the vicinity of the aerodrome, which may constitute a hazard to an aircraft; it's the controller's responsibility to deliver it with an easy identified manner.

Controller need to issue the information on essential local traffic in time, either directly or through approach controller when, in the judgment of the controller, such information is necessary in the interests of safety, or when requested by pilot.

4.3.3 Runway incursion or obstructed runway

In the event the controller observes, after a take-off clearance or a landing clearance has been issued, any obstruction on the runway likely to impair the safety of an aircraft taking off or landing, such as a runway incursion by an aircraft or vehicle, or animals or flocks of birds on the runway, he/she need to take the appropriate action as follows:

(a) in all cases inform pilot concerned of the obstruction and its location on the runway;

(b) cancel the take-off clearance for an aircraft which has not started to roll;

(c) instruct a landing aircraft to go around.

4.3.4 Abnormal aircraft configuration and conditions

The controller need to inform the aircraft concerned without delay whenever an abnormal configuration or condition of an aircraft, including conditions such as landing gear not extended or only partly extended, or unusual smoke emissions from any part of the aircraft.

When requested by the pilot of a departing aircraft suspecting damage to the aircraft, the departure runway used need to be inspected without delay and the controller shall advise the pilot as to whether any aircraft debris or bird or animal remains have been found or not.

4.4 Automatic terminal information service (ATIS)

To alleviate communication load on the ATS VHF air-ground channels and thus a consequent reduction on the workload of controller due to traffic density at some busy aerodromes, Automatic Terminal Information Service (ATIS) messages are broadcast to pass routine arrival/departure information on a discrete RTF frequency or on an appropriate VOR. The automatic terminal information service (ATIS) message is intended to provide the pilot with the complete range of information about an aerodrome necessary to allow him to make a definite decision about his approach and landing or his take-off.

Where ATIS broadcasting is used, the ATIS message may be combined into one broadcast, serving both arriving and departing traffic and, it may be split into one addressed specifically to arriving and one addressed to departing traffic. However, there are times when it is also suitable for controller to provide for individual direct transmission to pilot when the use of a broadcast is not warranted.

On first contact when aircraft inbound to aerodromes, aerodrome controller normally require pilot to inform him/her acknowledge receipt of current information by quoting the code letter of the broadcast. Outbound aircraft are not normally required to acknowledge receipt of departure ATIS except when requested on the actual ATIS broadcast. If, however, pilots report receipt of a departure ATIS broadcast the QNH are included thereby allowing controller to check that the quoted QNH is up-to-the-minute.

The hours of ATIS operation and the frequency employed are published in the AIP.

ATIS messages generally contain all or part of the following elements of information in the order listed:

(a) name of aerodrome;

(b) designator;

(c) time of observation, if appropriate;

(d) type of approach to be expected;

(e) the runway (s) in use; status of arresting system constituting a potential hazard, if any;

(f) significant runway surface conditions, and if appropriate, braking action;

(g) holding delay, if appropriate;

(h) transition level, if applicable;

(i) other essential operational information, if appropriate;

(j) surface wind direction and speed, including significant variations;

(k) visibility and, when applicable, RVR;

(l) present weather;

(m) cloud below 1 500 m (5 000 ft) or below the highest minimum sector altitude, whichever is greater; cumulonimbus; if the sky is obscured, vertical visibility when available;

(n) air temperature;

   (o) dew point temperature;

   (p) altimeter setting (s);

   (q) any available information on significant meteorological phenomena in the approach, take-off and climb-out areas;

   (r) trend-type landing forecast, when available;

   (s) specific ATIS instructions.

Exercise

I. Explain the following definitions.

(1) flooded   (2) slush   (3) VRB

(4) METAR/SPECI   (5) RVR   (6) OVC

(7) significant meteorological information   (8) runway incursion

II. Answer the following questions.

(1) What's the difference between CAVOK, SKC and NSC?

(2) What information elements are included in ATIS message?

(3) What related information is included in essential local traffic information?

III. Extract information from the following message.

(1) SPECI ZLXY 130044Z VRB01MPS 0800 R23/1200U FG NSC 13/13 Q1027 NOSIG =

(2) TAF ZYHB 190129Z 190312 04004MPS 3000 -RA BR SCT010 BKN030 BECMG 0405 01006G12MPS TEMPO 0509 1200 -RA - RASN SCT005 FEW023CB OVC023 =

Chapter 5

Separation in the vicinity of aerodrome

5. 1　General

In the vicinity of aerodrome, aerodrome controller is responsible for provision of vertical or horizontal separation:

(a) between all IFR flights;

(b) between IFR flights and special VFR flights;

(c) between IFR flights and VFR flights in Class C airspace;

(d) between special VFR flights.

Controller provides the separation minima by controlling the aircraft in the traffic circuit except that:

(a) aircraft in formation are exempted from the separation minima with respect to separation from other aircraft of the same flight;

(b) aircraft operating in different areas or different runways on aerodromes suitable for simultaneous landings or take-offs are exempted from the separation minima;

(c) separation minima does not apply to aircraft operating under military necessity.

Controller never gives clearance to execute any manoeuvre that would reduce the spacing between two aircraft less than the separation minimum applicable for that circumstance.

In additional, aerodrome traffic service unit may apply larger separations than the specified minima whenever exceptional circumstances such as unlawful interference or navigational difficulties call for extra precautions. This measure is done with due regard to all relevant factors so as to avoid impeding the flow of air traffic by the application of excessive separations. Unlawful interference with an aircraft constitutes a case of exceptional circumstances which might require the application of separations larger than the specified minima, between the aircraft being subjected to unlawful interference and other aircraft.

In case that the type of separation or minimum used to separate two aircraft cannot be maintained, another type of separation or minimum will be established prior to the time when the current separation minimum is infringed.

Whenever, as a result of failure or degradation of navigation, communications, altimetry, flight control or other systems, aircraft performance is degraded below the level required for the airspace in which it is operating. It is the controller's responsibility to take action to establish another appropriate type of separation or separation minimum upon the flight crew's request where the failure or degradation affects the separation minimum currently being employed.

In addition to the circumstances mentioned above, the separation minima may be reduced in the vicinity of aerodromes if:

(a) adequate separation can be provided by the aerodrome controller when each aircraft is continuously visible to this controller; or

(b) each aircraft is continuously visible to flight crews of the other aircraft concerned and the pilots thereof report that they can maintain their own separation; or

(c) in the case of one aircraft following another, the flight crew of the succeeding aircraft reports that the other aircraft is in sight and separation can be maintained.

5.2 Separation minimum between departing aircraft

Aerodrome controller need to take into account the following complementary provisions for the longitudinal separation minima between departing aircraft from an aerodrome.

(a) One-minute separation is required if aircraft are to fly on tracks diverging by at least 45 degrees immediately after take-off so that lateral separation is provided (see Fig 5.1). This minimum may be reduced when aircraft are using parallel runways.

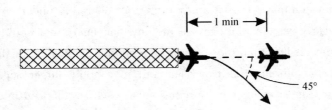

Fig 5.1 Separation between departing aircraft with tracks diverging more than 45°

(b) Two minutes are required between take-offs when the preceding aircraft is 74 km/h (40 kt) or faster than the following aircraft and both aircraft will follow the same track (see Fig 5.2).

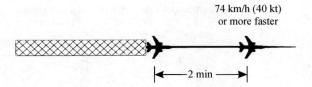

Fig 5. 2　Separation between aircraft following same track

（c）Five-minute separation is required while vertical separation does not exist if a departing aircraft fly through the level of a preceding departing aircraft and both aircraft propose to follow the same track（see Fig 5. 3）.

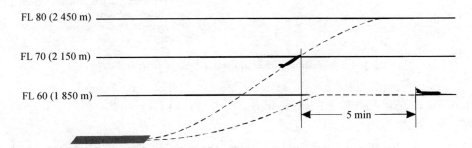

Fig 5. 3　Separation of departing aircraft following the same track

5. 3　Separation between departing aircraft from arriving aircraft

Controller applies the following separation when issues take-off clearance for a departing aircraft with consideration of the position of an arriving aircraft:

If an arriving aircraft is making a complete instrument approach, a departing aircraft may be cleared to take off:

（a）in any direction until an arriving aircraft has started its procedure turn or base turn leading to final approach;

（b）in a direction which is different by at least 45 degrees from the reciprocal of the direction of approach after the arriving aircraft has started procedure turn or base turn leading to final approach, provided that the takeoff will be made at least 3 minutes before the arriving aircraft is estimated to be over the beginning of the instrument runway（see Fig 5. 4）.

No take-offs in this area after procedure turn
is started nor within the last five minutes of a
straight-approach.

A. Straight-approach

45°

45°

B. Start of procedure turn

Take-offs pemitted in this area up to three
minutes before estimated arrival of aircraft
A or B or, in the case of A, until it crosses a
designated fix on the approach track.

Fig 5. 4   Separation of departing aircraft from arriving aircraft

If an arriving aircraft is making a straight-in approach, controller can clear a departing aircraft to take off:

(a) in any direction until 5 minutes before the arriving aircraft is estimated to be over the instrument runway;

(b) in a direction which is different by at least 45 degrees from the reciprocal of the direction of approach of the arriving aircraft:

1) until 3 minutes before the arriving aircraft is estimated to be over the beginning of the instrument runway (see Fig 5. 4), or

2) before the arriving aircraft crosses a designated fix on the approach track; the location of such fix to be determined by regulation authority after consultation with the operators.

5. 4   Wake turbulence separation

The term "wake turbulence" is used to describe the effect of the rotating air masses generated behind the wing tips of large jet aircraft. Pressure differential between the upward and downward of the wing triggers the rollup of the airflow aft of the wing resulting in swirling air masses trailing downstream of the wingtips. Vortices are generated from the moment an aircraft leaves the ground, since trailing vortices are the byproduct of wing lift (see Fig 5. 5). So generally it is unnecessary for the controller to consider the wake turbulence after a landing aircraft unless departing or landing after a large aircraft executing a low approach, missed approach, or touch and go landing (since vortices settle and move laterally near the ground, the vortex hazard may exist along the runway and in the flight path, particularly in a quartering tailwind), 2 minutes separation may be necessary prior to a takeoff or landing.

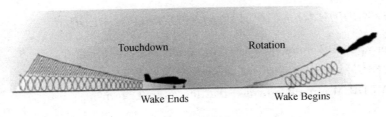

Fig 5.5　Vortex behavior

Controllers apply wake turbulence separation minima based on a grouping of aircraft types into three categories according to the maximum certificated take-off mass as followings:

(a) HEAVY (H) — all aircraft types of 136 000 kg or more;

(b) MEDIUM (M) — aircraft types less than 136 000 kg but more than 7 000 kg;

and

(c) LIGHT (L) — aircraft types of 7 000 kg or less.

5.4.1　Time-based wake turbulence separation

It's unnecessary for the aerodrome controller to apply wake turbulence separation:

(a) for arriving VFR flights landing on the same runway as a preceding landing HEAVY or MEDIUM aircraft; and

(b) between arriving IFR flights executing visual approach when the aircraft has reported the preceding aircraft in sight and has been instructed to follow and maintain own separation from that aircraft.

Aerodrome controller issues a caution of possible wake turbulence in respect of the flights specified above in (a) and (b), as well as when otherwise deemed necessary. Pilot is responsible for ensuring that the spacing from a preceding aircraft of a heavier wake turbulence category is acceptable. If pilot determine that additional spacing is required, he/she need to inform controller accordingly for this requirement.

5.4.1.1　Between arriving aircraft

Except the abovementioned circumstance (a) and (b), controller applies the following time-based separation minima to aircraft landing behind a HEAVY or a MEDIUM aircraft:

(a) MEDIUM aircraft behind HEAVY aircraft —2 minutes;

(b) LIGHT aircraft behind a HEAVY or MEDIUM aircraft— 3 minutes.

5. 4. 1. 2　Between departing aircraft

1. LIGHT/MEDIUM behind HEAVY or LIGHT behind a MEDIUM

Controller applies 2 minutes separation minimum (see Fig 5. 6, Fig 5. 7) between a LIGHT or MEDIUM aircraft taking off behind a HEAVY aircraft or a LIGHT aircraft taking off behind a MEDIUM aircraft provided that the aircraft are using:

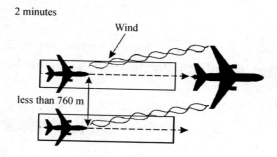

Fig 5. 6　Separation for following aircraft while using parallel runways

(a) the same runway;

(b) parallel runways separated by less than 760 m (2 500 ft);

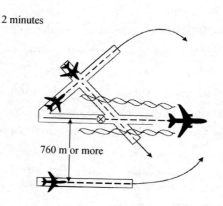

Fig 5. 7　Wake turbulence separation for crossing aircraft

(c)　crossing runways if the projected flight path of the second aircraft will cross the projected flight path of the first aircraft at the same altitude or less than 300 m （1 000 ft） below;

(d)　parallel runways separated by 760 m （2 500 ft） or more, if the projected flight path of the second aircraft will cross the projected flight path of the first aircraft at the same altitude or less than 300 m （1 000 ft） below.

2. LIGHT/MEDIUM behind HEAVY or LIGHT behind MEDIUM

Controller applies 3 minutes separation minimum （see Fig 5. 8） between a LIGHT or MEDIUM aircraft when taking off behind a HEAVY aircraft or a LIGHT aircraft when taking off behind a MEDIUM aircraft from:

(a)　an intermediate part of the same runway; or

(b)　an intermediate part of a parallel runway separated by less than 760 m （2 500ft） .

3 minutes

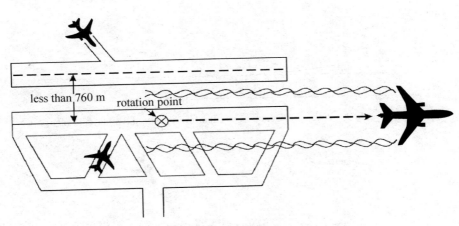

Fig 5. 8　Wake turbulence separation for following aircraft

3. Displaced landing threshold

Controller applies 2 minutes separation minimum between a LIGHT or MEDIUM aircraft and a HEAVY aircraft and between a LIGHT aircraft and a MEDIUM aircraft when operating on a runway with a displaced landing threshold when:

(a)　a departing LIGHT or MEDIUM aircraft follows a HEAVY aircraft arrival and a departing LIGHT aircraft follows a MEDIUM aircraft arrival; or

(b)　an arriving LIGHT or MEDIUM aircraft follows a HEAVY aircraft departure and an arriving LIGHT aircraft follows a MEDIUM aircraft departure if the projected flight paths are expected to cross.

4. Opposite direction

Controller applies 2 minutes separation minimum between a LIGHT or MEDIUM aircraft and a HEAVY aircraft and between a LIGHT aircraft and a MEDIUM aircraft when the heavier aircraft is making a low or missed approach and the lighter aircraft is:

(a) utilizing an opposite-direction runway for take-off (see Fig 5.9); or

(b) landing on the same runway in the opposite direction, or on a parallel opposite-direction runway separated by less than 760 m (2500 ft) (see Fig 5.10).

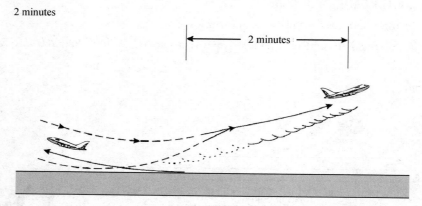

Fig 5.9   Wake turbulence separation for opposite direction take-off

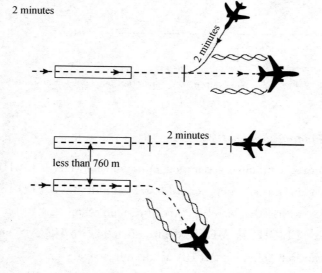

Fig 5.10   wake turbulence separation

5. 4. 2   Distance-based wake turbulence separation

Controller has to apply the following distance-based wake turbulence separation minima (see Table 5-1) to aircraft being provided with an ATS surveillance service in the approach and departure phases of flight in the circumstances given in.

Table 5-1   Radar Separation Minima difference between ICAO and CAAC

Aircraft category			
Preceding aircraft	Succeeding aircraft	Wake Turbulence Separation minima (ICAO)	Wake Turbulence Separation minima (China)
HEAVY	HEAVY	7. 4 km (4. 0 nm)	8 km
	MIDIUM	9. 3 km (5. 0 nm)	10 km
	LIGHT	11. 1 km (6. 0 nm)	12 km
MEDIUM	LIGHT	9. 3km (5. 0 nm)	10 km

The minima set out in above Table 5-1 are applied when:

(a) an aircraft is operating directly behind another aircraft at the same altitude or less than 300 m (1000 ft) below; or

(b) both aircraft are using the same runway, or parallel runways separated by less than 760 m; or

(c) an aircraft is crossing behind another aircraft, at the same altitude or less than 300 m (1000 ft) below.

5. 5   Visual separation

Visual separation is a kind of separation applied to aircraft authorized by the controller. Aircraft operating in airspace Class B, C, D or E may be separated by visual means and aircraft in airspace A will never be applied visual separation (generally up to 6000 meters) . Visual separation can reduce separation minima, relieve workload of controller and expedite traffic volume.

When applying visual separation, reported weather conditions must allow the aircraft to remain within sight until other separation established. Visual separation can not be applied between successive departures when departure routes and/or aircraft performance preclude maintaining separation. To assure other approved separation established before and after the application of visual separation, the following factors need to be considered:

（a） aircraft performance;

（b） wake turbulence;

（c） closure rate;

（d） routes of flight;

（e） known weather conditions.

5.5.1　Methods of visual separation application

Visual separation may be applied between aircraft under the control of the same ATS facility within the terminal area up to 6000 meters （including）. Air ground communication need to be maintained with at least one of the aircraft involved or the capability to communicate immediately. There are two methods to apply Visual applications between aircraft as followings.

5.5.1.1　Visual separation based on visual observation by aerodrome controller

Aerodrome controller can visually observe the aircraft and visual separation is maintained between the aircraft by controller. In this case, controller is responsible for the separation but cannot provide visual separation between aircraft when wake turbulence separation required or when the lead aircraft is a B757.

5.5.1.2　Visual separation based on visual observing another aircraft by the pilot

In this case, pilot sees another aircraft and maintains the separation from it so the pilot is responsible for the separation. Controller need only to instruct the pilot to maintain visual separation with the following procedure:

（a） Inform pilot about the other aircraft including position, direction and, unless it is obvious, the other aircraft's intention;

（b） Obtain acknowledgment from the pilot that the other aircraft is in sight;

（c） Instruct the pilot to maintain visual separation from that aircraft;

（d） Advise the pilot if the radar targets appear likely to converge;

（e） If the aircraft are on converging courses, inform the other aircraft of the traffic and that visual separation is being applied;

（f） If the pilot advises he/she has the traffic in sight and will maintain visual separation from it （the pilot must use that entire phrase）, the controller need only "approve" the operation instead of restating the instructions.

5. 6   Runway separation

5. 6. 1   Separations on the same runway

To separate a departing or landing aircraft from a preceding departing or arriving aircraft using the same runway (see Fig 5. 11, Fig 5. 12), aerodrome controller has to ensure that he/she will not permit pilot to commence take-off or cross the runway threshold on final approach until:

(a) the preceding departing aircraft has crossed the end of the runway-in-use or has started a turn;

(b) all preceding landing aircraft are clear of the runway-in-use.

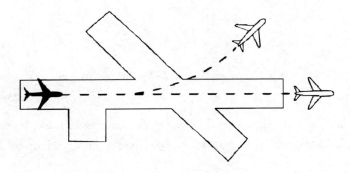

Fig 5. 11   Same runway separation

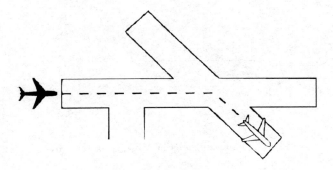

Fig 5. 12   Preceding landing aircraft clear of runway

5.6.2　Separation on the intersecting runway

To separate departing aircraft from an aircraft using an intersecting runway, or nonintersecting runways but the flight paths intersect, aerodrome need to ensure that the departure does not begin takeoff roll until one of the following circumstances:

(a) The preceding aircraft has departed and passed the intersection, has crossed the departure runway, or is turning to avert any conflict (see Fig 5.13, Fig 5.14).

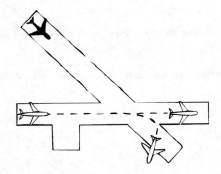

Fig 5.13　Separation on intersecting runway (a)

(b) A preceding arriving aircraft is clear of the landing runway, completed the landing roll and will hold short of the intersection, passed the intersection, or has crossed over the departure runway (see Fig 5.15, Fig 5.16).

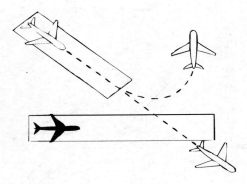

Fig 5.14　Separation on intersecting runway (b)

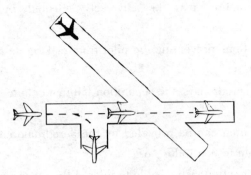

Fig 5. 15   Separation on intersecting runway （c）

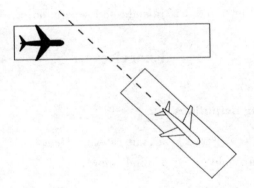

Fig 5. 16   Separation on intersecting runway （d）

（c）Wake turbulence application

5. 7 Reduction of separation minima

Provided an appropriate safety assessment has shown that an acceptable level of safety has been met, lower minima than abovementioned may be applied, after consultation with the operators, and taking into account such factors as:

（a）runway length;

（b）aerodrome layout; and

（c）types of aircraft involved.

But this lower minima will not be applied:

1) between a departing aircraft and a preceding landing aircraft;

2) between sunset and sunrise, or such other period between sunset and sunrise as

may be prescribed;

 3) when braking action may be adversely affected by runway contaminants (e. g. slush, water, etc.); and

 4) in weather conditions preventing the pilot from making an early assessment of traffic conditions on the runway.

 In the vicinity of an aerodrome, the separation minima detailed may be reduced by the aerodrome controller if:

 (a) adequate separation can be provided by the aerodrome controller when each aircraft is continuously visible to controller; or

 (b) each aircraft is continuously visible to pilot of the other aircraft concerned and the pilots thereof report that they can maintain their own separation; or

 (c) in the case of one aircraft following another, the flight crew of the succeeding aircraft reports that the other aircraft is in sight and separation can be maintained.

Exercise

I . Explain the following definitions.

(1) wake turbulence   (2) aircraft category Heavy   (3) visual separation

(4) displaced runway threshold   (5) parallel runway   (6) OVC

II . Answer the following questions.

(1) In the vicinity of aerodromes, what kinds of flight are provided vertical or horizontal separation?

(2) How many minutes separation is provided when a departing aircraft fly through the level of a preceding departing aircraft and both aircraft propose to follow the same track while vertical separation does not exist ?

(3) If an ATS authority attempts to reduce the separation minima, what kind of factors shall be taken into account?

(4) When separate a departing aircraft from a preceding departing or succeeding arriving aircraft using the same runway, under what conditions can a controller clear the departing aircraft to commence take-off run and the landing aircraft to cross the runway threshold ?

(5) What's the separation minimum between a MEDIUM and HEAVY aircraft when the heavier aircraft is making a low or missed approach and the lighter aircraft is using an opposite-direction runway for take-off?

(6) If an arriving aircraft is making a complete instrument approach, what's the premise that the take-off clearance for a departing aircraft may be issued?

(7) What's the wake turbulence separation between the HEAVY departing aircraft and HEAVY arriving aircraft? and why?

Chapter 6

Aerodrome ATS operation procedure

6. 1 Responsibility of aerodrome control service

ATC service is sub-divided into three parts, depending on the stage of flight to which it is performed. These three services are identified as Area Control Service, Approach control service, and Aerodrome control service.

At and in the vicinity of aerodromes, air traffic control service is normally provided by the aerodrome control tower. Aerodrome control service provides the air traffic control service for aerodrome traffic, except for those parts of controlled flights associated with arrival or departure, in order to accomplish objectives to prevent collisions between aircraft, to prevent collisions between aircraft on the maneuvering area and obstructions on that area, and to expedite and maintain an orderly flow of air traffic.

The need for the provision of aerodrome air traffic service is determined by consideration of the following:
(a) the types of air traffic involved;
(b) the density of air traffic;
(c) the meteorological conditions;
(d) such other factors as may be relevant.

Approach control service (APP) is also provided in the vicinity of aerodromes, but is a service which is mainly concerned with flights operating on an instrument flight rules (IFR).

The division of responsibilities between TWR and APP and between APP and ACC cannot be rigidly defined because the responsibilities depend very much on local conditions which vary from location to location. They must therefore be determined in each case and with due regard to traffic conditions, its composition, the airspace arrangements, prevailing meteorological conditions and relative workload factors. However, arrangements governing the division of responsibilities between these different parts of the ATS service, should not result in increased requirements for co-ordination and/or an undesirable inflexibility in the use of airspace, nor in an increased workload for both pilots and controller because of unnecessary transfers of control and associated radio communication contacts.

The establishment of an aerodrome control service does not necessarily imply the immediate provision of a special ATC facility (control tower) but it is rather intended to mean that the service will be provided by adequately qualified ATC personnel, having means and facilities at their disposal appropriate for the given situation. These means and facilities can range from relatively simple arrangements to a complete system of ATC services, including radio voice communication and electronic data processing and display equipment.

6.2   Aerodrome airspace

6.2.1   Airspace responsible by aerodrome control tower

According to relevant ICAO regulations, ATS airspaces are classified and designated as Class A, B, C, D, E, F and G. Among them, Airspace Classes A to E is controlled airspace. Controlled airspace is the airspace of defined dimensions within which air traffic control service is provided in accordance with the airspace classification.

The responsible airspace by an aerodrome control tower, in addition to aerodrome traffic, generally considers all traffic operating within a reasonable distance of the aerodrome. While no precise limitations can be imposed because the distance will vary in accordance with the traffic handled at the moment, generally within 25 NM radiuses.

Where there is only VFR traffic at the aerodrome, the designation of a controlled airspace is not necessary and also not generally desirable*. However, if the density of VFR traffic reaches proportions which could make the traffic pattern of departing and arriving aircraft difficult because of 'overflying' aircraft, an aerodrome traffic zone may be established to permit the control tower either to exercise control over aircraft not intending to land at the aerodrome in question or to make them avoid that zone.

Aerodrome traffic zone may also be established around uncontrolled aerodromes when the activities conducted at those aerodromes (i. e. flying school, specific military activities) make it undesirable for other aircraft, not engaged in these activities, to penetrate or otherwise disturb the traffic pattern*. In this case, the aerodrome traffic zone is primarily reserved for use by aircraft participating in the activities having caused the zone to be established.

Note: * *denotes FAA regulation, the same in other chapters.*

6.2.2　Control zone

Control zone is a controlled airspace extending upwards from the surface of the earth to a specified upper limit. When further developments determine that it is necessary for traffic operating under IFR, it will be necessary to protect such traffic by extending control to those traffic imposing restrictions on VFR flights are necessary to ensure the safety of both types of operations while operating in the same general area. To accomplish this, sufficient controlled airspace should be established to encompass the arrival, departure and, where necessary, the holding flight paths of the IFR flights. To achieve this in the most efficient manner, it will generally suffice to establish a comparatively small control zone (which, by definition extends from the ground upwards) and to superimpose on it a control area of a size sufficient to contain the flight paths of departing, holding and arriving IFR flights. In doing so, the lateral extent of the control zone is determined in relation to the lower limit of the superimposed control area so that average flight trajectories during departure and arrival are fully contained within the totality of the controlled airspace formed by the control zone and the control area. In addition, unnecessary restrictions are not imposed on other VFR air traffic wishing to operate in airspace close to the controlled aerodrome but not wishing to use that aerodrome itself. Aerodrome control zone attributes to Class airspace D.

The lateral limits of a control zone generally extend to at least 9.3 km (5 NM) from the centre of the aerodrome or aerodromes concerned in the directions to make approach, control zone may include two or more aerodromes situated close together.

6.2.2.1　Airspace structure in China

In China, according to the functions to be performed, the airspaces are classified as upper level control area, low level control area, terminal (approach) control area and aerodrome control area respectively. They are all controlled airspace and according to the intention, they are divided as aerodrome flight airspace, airways, air routes, prohibited areas, restricted areas and danger areas. Air corridors, fuel dumping areas and temporary flight airspace may be established when necessary for the need of airspace management and flight missions. Aerodrome flight airspace is established off airways and air corridors. The horizontal separation between the limits of instrument (in clouds) flight airspace and those of the airways, air corridors and other airspace shall be not less than 10 kilometers.

Normally aerodrome flight airspace includes such flight airspace as flying techniques (aerobatic, formation, instrument) flight airspace, scientific research test flight airspace, firing flight airspace, low-level flight airspace, super-low-level flight airspace, aeromarine flight airspace, nighttime flight airspace and holding airspace.

The establishment of airspace takes into consideration such factors as the national security, the flight requirements, the capability of flight control, the availability of communication, navigation and radar facilities and the locations of aerodrome and environment protection.

6. 2. 2. 2　Aerodrome control zone in China

The lateral limit of aerodrome control zone in China is designated as following:

(a) to aerodrome which is capable of departing and landing aircraft Category D or above. The lateral limits of control zones is the area enclosed by two arcs which center locates the thresholds of each end of the runway respectively and radius 13km, and two lines at each side of the runway which are parallel with the runway and abeam the arcs;

(b) to aerodrome which is only capable of departing and landing aircraft Category C or below. The specification of the lateral limits of control zone is the same with (a), except that the radius of the arcs or the circle centered at the datum mark is 10 km;

(c) to aerodrome which is only for departing and landing aircraft Category B or below. The lateral limits of control zones is the circle which centered at the datum mark and radius 10 km;

(d) to those aerodrome which needs special approach procedure operation. The lateral limits of control zones can be properly enlarged according to practical needs.

6. 3　Functions of aerodrome control tower

The functions of aerodrome control tower are to issue information and clearances to aircraft under their control to achieve a safe, orderly and expeditious flow of air traffic in the vicinity of an aerodrome with the object of preventing collision (s) between:

(a) Aircraft flying within the designated area of responsibility of the control tower, including the aerodrome traffic circuits;

(b) Aircraft operating on the manoeuvring area;

(c) Aircraft landing and taking off;

(d) Aircraft and vehicles operating on the manoeuvring area;

(e) Aircraft on the manoeuvring area and obstructions on that area.

Aerodrome controllers maintain a continuous watch on all flight operations at and in the vicinity of an aerodrome as well as vehicles and personnel on the manoeuvring area. Controller maintains watch by visual observation, augmented in low visibility conditions by radar if it is available. Controller controls the traffic in accordance with the procedures and all applicable traffic rules. If there are other aerodromes within the control zone, traffic at all aerodromes within the zone is coordinated so that traffic circuits do not conflict.

The functions of an aerodrome control tower may be performed by different control or working positions, such as:

(a) Aerodrome controller, normally responsible for operations on the runway and aircraft flying within its responsible airspace;

(b) Ground controller, normally responsible for traffic on the manoeuvring area with the exception of runways;

(c) Clearance delivery position, normally responsible for delivery of start-up and ATC clearances to departing IFR flights.

Where parallel or near-parallel runways are used for simultaneous operations, individual aerodrome controllers are responsible for operations on each of the runways.

Aerodrome control towers are also responsible for alerting the rescue and fire fighting services whenever:

(a) An aircraft accident has occurred on or in the vicinity of the aerodrome; or,

(b) Information is received that the safety of an aircraft is or will come under the jurisdiction of the aerodrome control tower may have or has been impaired; or

(c) Requested by the pilot; or

(d) When otherwise deemed necessary or desirable.

An aerodrome control tower has two major operational requirements to be able to properly control aircraft operating on and in the vicinity of the aerodrome. Those requirements are:

(a) The tower must permit the controller to survey those portions of the aerodrome and its vicinity over which he exercises control;

(b) The tower must be equipped so as to permit the controller rapid and reliable communications with aircraft with which he is concerned.

Surveillance by the aerodrome controller is normally done by visual means (eyesight) alone, mechanically through the use of binoculars to improve eyesight or electronically, through the use of ATS surveillance system or closed-circuit television. The controller must be able to discriminate between aircraft and between aircraft and vehicles while they are on the same or different runways and/or taxiways.

The most significant factors contributing to adequate visual surveillance are the site of the tower and the height of the control tower cab. The optimum tower site normally is as close as possible to the centre of the maneuvering part of the aerodrome, provided that at the intended height, the tower structure itself does not become an obstruction or hazard to flight.

6. 4　Runway selection

The term "runway-in-use" is used to indicate the runway or runways that, at a particular time, are considered by the aerodrome controller to be the most suitable for use by the type of aircraft expected to land or take off at the aerodrome. Separate or multiple runways may be designated runway-in-use for arriving aircraft and departing aircraft.

Normally, an aircraft lands and takes off into wind unless safety, the runway configuration, weather conditions and available instrument approach procedures or air traffic conditions determine that a different direction is preferable. In selecting the runway-in-use, however, controller needs to take into consideration, besides surface wind speed and direction, other relevant factors such as the aerodrome traffic circuits, the length of runways, and the approach and landing aids available.

Generally, aerodrome controller select the runway most nearly aligned with the upwind when it's 5 knots or more, or select the "calm wind" runway when less than 5 knots unless use of another runway.

If the runway-in-use is not considered suitable for the operation involved, the pilot may request for controller's permission to use another runway and, if circumstances permit, controller should clear that request accordingly.

STOL runways may be used only when requested by the pilot or as specified in a letter of agreement with an aircraft operator and the aerodrome controller issue the measured STOL runway length if the pilot requests it.

SUMMARY:
Controller considers the following factors when deciding Runway in use:
- Surface wind speed and direction;
- Aerodrome traffic circuits;
- Length of Runways;
- Approach & landing aids available.

6. 5　Flight operation procedure in the vicinity of aerodrome

6. 5. 1　Departure procedure

The following departure procedures are based on those applicable for an aerodrome that have all available services, and are listed in the order that they would be used. At smaller, less equipped aerodromes, some services are combined, e. g. , the IFR clearance is obtained from ground control where there is no separate clearance delivery frequency.

6. 5. 1. 1　ATIS broadcasts

Pilot can obtain the ATIS information prior to contacting either the ground control or tower if ATIS is available.

6. 5. 1. 2　Clearance delivery

Prior to requesting taxi authorization from controller, pilot of IFR departures has to call on clearance delivery frequency, normally no more than 5 minutes before engine starts. Where a clearance delivery frequency is not published in AIP, the IFR clearance is normally given after taxi authorization. At some major aerodromes, departing VFR aircraft may also be required to contact "clearance delivery" before taxiing. In China, all departures have to get clearance.

6. 5. 1. 3　Requests for push-back or power-back

At some aerodromes, controllers are not be in a position to see all obstructions aircraft may encounter during push-back or power-back. Under this circumstance, it's pilot's responsibility to ensure that push-back or power-back is accomplished safely prior to initiating aircraft movement.

6. 5. 1. 4　Taxi information

Controller delivers taxi authorization to pilot on ground control frequency. Upon receipt of a normal taxi clearance, pilot is expected to proceed to the taxi-holding position for the runway assigned for takeoff. If controller requires the pilot to cross any runway while taxiing towards the departure runway, he/she shall issue a specific instruction to cross or hold short. If this specific authorization to cross was not received, pilots need to hold short and request clearance to cross the runway.

Detailed taxi information will be given in the following sections.

6. 5. 1. 5　Taxi holding positions

Clearance from controller must be obtained before leaving a taxi holding position, or where a holding position marking is not visible or has not been established.

6.5.1.6　Taxiway holding positions during IFR operations

At controlled aerodrome, it is imperative that aircraft waits beyond taxiway holding signs until cleared by controller. Aircraft proceeding beyond the taxiway holding position signs may infringe sensitive areas and cause dangerous interference to the glide path or localizer signals.

When an aerodrome is operating under CAT II/III weather conditions or when its CAT II/III operations plan is in effect, controller need to follow the specific procedure and pilots are to observe CAT II or III mandatory holding position signs.

6.5.1.7　Take-off clearance

When ready for takeoff, pilot needs to request the take-off clearance with the runway number from the controller. Upon receipt of the clearance, the pilot also needs to acknowledge and take off without delay, or inform the controller if unable to do so.

6.5.1.8　Release from tower frequency

Unless otherwise advised by the controller, normally pilot doesn't need to require permission to change from tower frequency once clear of the control zone. He/she doesn't need to request release from this frequency or report clear of the zone when there is considerable frequency congestion.

6.5.2　Arrival procedures

Prior to contacting aerodrome controller, all arrivals need to monitor the frequency to obtain the basic aerodrome information if ATIS is available. Contents of ATIS will be given in the following sections.

6.5.2.1　Initial contact

Pilots need to establish and maintain radio communications with aerodrome controller prior to operating within any control zone. It is recommended that the pilot contact the tower about 5 minutes prior to requiring clearance or entering the zone.

6.5.2.2　Initial clearance

On initial contact with aerodrome controller, unless the pilot advises receipt of ATIS, controller need to inform pilot runway in use, wind direction and speed, altimeter setting and other pertinent information. After that, controller issues clearance to proceed, including any necessary restrictions.

6. 5. 2. 3   VFR holding procedures

Aerodrome controller may ask the pilot to ORBIT visually over a geographic location, VFR checkpoint or call-up point when it is required by traffic until clear them to the aerodrome. If the request is not acceptable, pilots can inform the controller and state their intentions.

6. 5. 2. 4   Landing clearance

At controlled aerodromes, pilot needs to obtain landing clearance from aerodrome controller prior to landing. Normally, controller initiates landing clearance without having received the request from the pilot. Landing clearance is normally given when aircraft is on downwind or final approach. If landing clearance is not received, the pilot must pull up and make another circuit except in case of emergency.

6. 5. 2. 5   Taxiing

Pilot needs to obtain controller clearance to taxi on the maneuvering area at a controlled aerodrome. Unless otherwise instructed by the controller, aircraft are expected to continue in the landing direction to the nearest suitable taxiway, exit the runway without delay and request further taxi instructions. After landing on a dead-end runway, controller gives instructions to backtrack.

6. 5. 3   Aerodrome traffic circuit

6. 5. 3. 1   Purpose of the aerodrome traffic circuit

Traffic circuit can be considered to be a special pattern, or a series of paths used by VFR traffic to fly to and away from the runway in use in the vicinity of an aerodrome. The primary purpose of the traffic circuit is to standardize the methods of approaching a runway to land, so that pilot is aware of the likely position of other aircraft in the vicinity of an aerodrome. Normally traffic circuit is left hand and has a rectangular pattern. It begins and ends over the runway and is generally flown between 500 to 1500 feet above the airport elevation. It may be right hand under some circumstances such as terrain, noise-sensitive area require all turns in the aerodrome traffic circuit to be made to the right.

6. 5. 3. 2　Traffic circuit components

Traffic circuit consists of the upwind leg, crosswind leg, downwind leg, base leg and final approach leg which are illustrated (see Fig 6. 1) as followings:

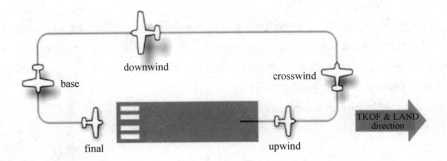

Fig 6. 1　Traffic circuit

The upwind leg begins at the point where the aircraft leaves the ground. It continues climbing straight ahead to gain the sufficient altitude before the 90-degree left turn is made to the crosswind leg.

The crosswind leg is a flight path at a 90° angle to the takeoff direction. After making a left turn from the upwind leg one enters the crosswind leg. This turn is made at a safe height, while the climb is continued towards the indicated or cleared circuit altitude.

The downwind leg is a flight path parallel to the landing runway in the opposite of the landing direction with the runway at the left side of the aircraft.

The base leg is a flight path at a 90° angle to the landing runway direction and connects the downwind leg to the final approach leg.

The final approach leg is a flight path in the direction of landing from the base leg to the runway.

6. 5. 3. 3　Traffic circuit entry procedure

When aerodrome controller gives a clearance "to the circuit" for aircraft, it is expected that the aircraft joins the circuit on the downwind leg at circuit height. Depending on the direction of approach to the aerodrome and the runway in use, it may be necessary to proceed crosswind prior to joining the circuit on downwind leg.

A straight-in approach is an approach where an aircraft joins the traffic circuit on the final without having executed any other portion of the circuit.

Aerodrome controller may clear a right-hand approach while a left-hand circuit is in effect. It is expected that the aircraft will join the circuit on the right-hand downwind leg, or join directly into the right-hand base leg.

The aircraft entry to the circuit need to avoid cutting off other aircraft, conforming as closely as possible to the altitude (normally 1000 ft), speed and size of the circuit being flown by other traffic.

In order to increase safety by reducing the possibility of conflicting with departing traffic, aircraft approaching the active runway from the upwind side are to join the downwind leg abeam a point approximately midway between each end of the runway, taking into account aircraft performance, wind and/or runway length.

A typical traffic circuit entry procedure is demonstrated as Fig 6.2.

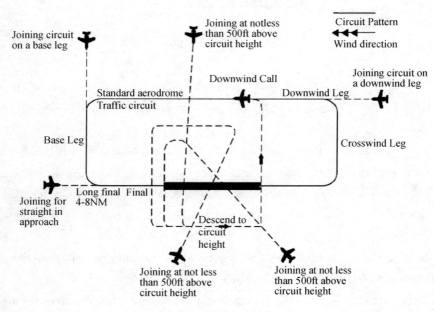

Fig 6.2   Traffic circuit entry procedure

Traffic circuit altitude normally is 300 to 500 meters.

6.5.4   Visual approach

Visual approach is an ATC authorization for an aircraft on an IFR flight plan to proceed visually to the aerodrome of intended landing; it is not an instrument approach procedure, so there is no missed approach segment. An aircraft which is unable to complete a visual approach will be handled as go-around and appropriate separation need to be provided by controller.

6. 5. 4. 1 Visual approach weather condition requirements

Reported ceiling at the aerodrome of intended landing is at least 300 meters or greater while the visibility is 5 kilometers or greater. At airports without weather reporting service there must be reasonable assurance that descent flight to the aerodrome can be made visually while the pilot must be informed that weather information is not available.

Controller may initiate radar vector for visual approach if reported ceiling is 150 meters above the minimum radar vectoring altitude and the visibility is 5 kilometers or greater.

6. 5. 4. 2 Visual approach procedure

Visual approach may be initiated by controllers or pilots request even when an aircraft is being vectored for an instrument approach. When controller initiates visual approach, the concurrence of the flight crew is needed.

Visual approach can be applied as the following procedures:

(a) Pilot reports aerodrome or the runway in sight;

(b) Resolve potential conflicts with all other aircraft, advise an overtaking aircraft of the distance to the preceding aircraft and speed difference, and ensure that weather conditions at the aerodrome are VFR or that the pilot has been informed that weather is not available for the destination aerodrome;

(c) All aircraft following a heavy jet/B757 must be informed of the airplane manufacturer and model;

(d) In those instances where aerodromes are located in close proximity, the locations of the aerodrome that may cause the confusion need to be provided;

(e)* The pilot reports the aerodrome or runway in sight but not the preceding aircraft. Radar separation must be maintained until visual separation is provided.

Separation need to be provided between aircraft cleared to execute a visual approach and the other arriving and departing aircraft. For successive visual approaches, controller maintains radar or non-radar separation until the pilot of a succeeding aircraft reports having the preceding aircraft in sight. The aircraft is then be instructed to follow and maintain own separation from the preceding aircraft. When both aircraft are heavy wake turbulence category or the preceding aircraft is a heavier wake turbulence category than the following, and the distance between the aircraft is less than the appropriate wake turbulence minimum, the controller need to issue a caution of possible wake turbulence.

6. 5. 4. 3　Visual approach clearance

The controller may clear an aircraft to make a visual approach when:

(a)　The aircraft is number one in the approach sequence;

(b)　The aircraft is to follow a preceding aircraft and the pilot reports the preceding aircraft in sight and is instructed to follow it.

6. 6　VFR flights operations in the vicinity of aerodrome

6. 6. 1　Suspension of VFR operations

Any or all VFR operations on and in the vicinity of an aerodrome may be suspended by any of the following units, or authorities whenever safety requires such action:

(a)　approach control unit or appropriate ACC;

(b)　aerodrome control tower;

(c)　ATS authority.

Suspensions of VFR operations generally are accomplished through or notified to the aerodrome control tower.

Aerodrome controller follows the following procedures whenever VFR operations are suspended:

(a)　hold all VFR departures;

(b)　recall all local flights operating under VFR or obtain approval for special VFR operations;

(c)　notify the approach control or ACC as appropriate of the action taken;

(d)　notify all operators, or their designated representatives, of the reason for taking such action, if necessary or requested.

6. 6. 2　Authorization of special VFR flights *

Special VFR flight is a VFR flight cleared by ATC to operate within a control zone in meteorological conditions below VMC. Currently there is no special VFR flights operation in China.

When the ground visibility is not less than 1500 m, and the traffic conditions permit, special VFR flights may be authorized subject to the approval of the unit providing approach control service. The special VFR flights may be authorized to enter a control zone for the purpose of landing, take off and depart from a control zone, cross a control zone or operate locally within a control zone.

Requests of every flight for such authorization are handled individually; separation is effected between all IFR flights and special VFR flights in accordance with separation minima in Chapter 5, when it's prescribed by ATS authority, all special VFR flights in accordance with separation minima prescribed by that authority.

6. 7　Surface movement guidance and control

In its broadest sense, surface movement guidance and control (SMGC) is the provision of guidance to, and the control of, all aircraft and ground vehicles on the movement area of an aerodrome. "Guidance" relates to equipment, information and advice necessary to enable pilots or the drivers of ground vehicles to find their way on the aerodrome and to keep the aircraft or vehicles on the surfaces or within the areas intended for their use. "Control" means the measures necessary to prevent collisions and to ensure that the traffic flows smoothly and efficiently.

SMGC system provides guidance to, and control of, an aircraft from the landing runway to the parking position on the apron and back again to the runway used for take-off, as well as from the maintenance area to the apron, or vice versa. The system also provides guidance to, and control of, all ground vehicles whose functions require them to operate on the movement area, e. g. aerodrome management vehicles, aircraft servicing vehicles, rescue and fire fighting vehicles, and vehicles engaged in construction work. Additionally, SMGC system assists in safeguarding against unauthorized or inadvertent entry onto operational runways.

SMGC system comprises an appropriate combination of visual aids, non-visual aids, radiotelephony communications, procedures, control and information facilities. Systems range from very simple arrangements at small aerodromes with light traffic operating only in good visibility to very complex facilities at large and busy aerodromes with operations in very low visibility conditions.

At some aerodromes, the task of providing SMGCS has been assigned to the aerodrome control tower, the operational responsibility for co-ordinating the movement of aircraft on the movement area belongs to relevant ATS unit. Whereas some aerodromes, it's aerodrome authority's responsibility to provide SMGC guidance.

ATS unit and/or aerodrome authority need to monitor the SMGC system and to have any failures corrected as soon as possible. The monitoring may take the form of visual surveillance of lights, including reports from pilots, and of electronic monitoring of the electrical and electronic components of the system.

6.7.1　Apron management service

The provision of surface movement control service on the maneuvering area of an aerodrome is within the specifications concerning aerodrome control service. Apron management service is a service provided to regulate the activities and the movement of aircraft and vehicles on the apron. With regard to the management of aircraft and vehicles on the apron, a variety of different methods are available depending on the particular conditions at an aerodrome. Apron management service may be provided by the aerodrome ATS unit, or by the aerodrome authority, or by the operator in the case of a company terminal, or by co-ordinated service between ATS and the aerodrome authority or operating company. Details concerning the different possible methods for the provision of the apron management service are included in the Manual of SMGC Systems.

At certain aerodromes, control of aircraft on the apron is not the responsibility of ATS units. At these aerodromes, there should be a designated body responsible for ensuring the safe movement of aircraft on the apron.

6.7.2　Surface movement radar （SMR）

Surface movement radar （SMR） has proven to be useful in assisting with the monitoring of aircraft and vehicles on the maneuvering area. It's one of the possible components of an SMGC system and may be used as an aid in the provision of aerodrome control service.

SMR, like primary radar, is a stand-alone type system and does not warrant the technical definition or co-ordination required by technically or operationally interdependent systems. Considering that each surface movement guidance and control system must be related to the operational conditions and requirements of the particular aerodrome （i. e. visibility conditions, traffic density and aerodrome layout）, the system composition and capability are considered matters to be decided on an individual basis. SMR has proven to be useful in assisting with the monitoring of aircraft and vehicles on the maneuvering area particularly in low visibility conditions.

There is currently no facility, or combination of facilities, that compensates fully for a controller's loss of visual contact with the aerodrome surface traffic. Information derived by other methods such as radiotelephony （RTF） communications or SMR is rarely comprehensive, and is far less economical in terms of the workload required for its acquisition. Due to the normally distant location of an aerodrome control tower from runways and taxiways, the controller's ability to control traffic on the maneuvering area on the basis of visual observation will, during periods of reduced visibility, be limited. When such conditions of reduced visibility prevail, the air traffic control (ATC) workload per movement increases and the traffic handling capacity of the aerodrome control service may be reduced.

Provided that an aerodrome is adequately equipped with visual aids, the provision of an aerodrome SMR can make a valuable contribution to the safety and efficiency of the aerodrome surface movement control during periods of low visibility and at night; optimum capacity under these conditions is unlikely to be achieved without it. SMR permits a continuous check of runway occupancy and taxiway usage. Furthermore, it allows rapid appreciation of lighting control requirements and facilitates clearances for aircraft and vehicles. In emergencies it can play an important part in expediting movement of emergency vehicles and the safe disposition of other traffic.

At a major aerodrome, a large part of the maneuvering area may be obscured at times from the aerodrome control tower while visibility is still within the limits at which traffic can operate at the normal level of demand, i. e. in visibility conditions sufficient for the pilot to taxi and avoid collision with other traffic on taxiways and at intersections by visual reference, but insufficient for aerodrome controller to exercise control and detect runway intrusions on the basis of visual observation. The introduction of SMR can to a great degree alleviate the limitations on observation and control normally associated with low visibility conditions; however, the workload involved in detailed monitoring, together with the clutter and other limiting factors, is very high and may therefore restrict the number of movements the aerodrome controller can handle at one time. Obviously, the accuracy of maneuvers required on taxiways which can be satisfactorily accomplished by following lights and markings is far more precise than could be provided by ATC instructions using SMR guidance. It is necessary for the pilot to be able to comply with instructions given by the controller without the radar being used to provide directional guidance. However, the traffic and positional information a controller is able to provide by using SMR where it is provided as a component of the surface movement guidance and control (SMGC) systems is of major assistance to pilots operating on the manoeuvring area of an aerodrome. SMR can make a valuable contribution to the safety and efficiency of aerodrome surface movement control in low visibility conditions and at night. However, it should be emphasized that SMR is an adjunct and not an alternative to the visual aids and procedures currently used for the control of aircraft and vehicles on the manoeuvring area.

6. 7. 2. 1　Use of SMR

Aerodrome controllers determine the position of aircraft and vehicles on the manoeuvring area by visual observation and/or radio position reports. Within the limitation of the radar coverage, the information presented on an aerodrome SMR display may be used by aerodrome controller to supplement these existing methods as follows:

(a) to confirm that the runway is clear of aircraft, vehicles or obstructions prior to a departure or landing;

(b) to ensure that the departing aircraft is lined up on the correct runway;

(c) to ensure that the arriving aircraft has vacated the runway;

(d) to ascertain that the departing aircraft has commenced take-off run;

(e) to provide directional information to pilots or vehicle operators on request or as necessary;

(f) to monitor aircraft/vehicle compliance with control instructions on the manoeuvring area;

(g) to monitor the manoeuvring area and identify optimum taxiing routes that reduce congestion and assist in expediting the low of traffic during periods of low visibility;

(h) to confirm a pilot or vehicle operator position report; to provide guidance information to emergency vehicles, as necessary;

(i) to assist in the timing of landing and take-off clearances in low visibility conditions to maximize runway utilization;

(j) to provide detection and guidance information to an aircraft uncertain of its position;

(k) to assist in detecting runway intrusions; and to ensure that approving of requested push-back will not conflict with traffic on the maneuvering area.

SMR may be used to assist in conflict resolution at intersections and as an aid to assignment of intersection priorities where a possible conflict exists. This function is performed by the issuance of appropriate holding instructions. Observation of the general traffic pattern and points of congestion on an SMR display provides information to assist the controller in determining which aircraft should be given priority at an intersection.

SMR may be used to ensure that a runway is clear of traffic before clearance is given for a landing or take-off on that runway. An arriving aircraft leaving the runway report "runway vacated" on the appropriate frequency. During periods of low visibility, however, it is sometimes difficult for the pilot to confirm that the aircraft is clear of the runway in use. SMR may, therefore, be used to verify a "runway vacated" report from the pilot. An aircraft which is approaching a runway on an intersecting taxiway and has been given an instruction to hold should be monitored on the SMR to confirm compliance. Intersecting runways may be monitored on the SMR for possible conflicts prior to clearing aircraft for take-off or landing.

SMR may also be used to ensure that a departing aircraft has taxied into position for take-off on the proper runway. Such a check is particularly important when two close parallel runways are in use and an arriving aircraft is on final approach for landing on the adjacent runway.

Runway utilization during periods of low visibility can be significantly improved through the use of SMR for those runway configurations that involve interaction of arriving and departing traffic during take-off or landing. Anticipation of the runway turn-off by an arriving aircraft can be gained through SMR. Compliance with "line up and wait" instructions to departing aircraft, commencement of take-off roll and lift-off can be monitored on SMR. Use of SMR can significantly assist in seizing opportunities for the release of departing aircraft between arriving aircraft during low visibility conditions whenever runways are used for both arrivals and departures.

6. 7. 2. 2   Limitations of SMR

Any of the following technical limitations affects the operational efficiency and use of SMR:

(a) aircraft/vehicle size-detectability diminishes with reduction in size;

(b) line-of-sight limitations;

(c) heavy rain causing clutter and resolution difficulties;

(d) shielding - a portion of an aircraft/vehicle may be shielded from the radar by another part of the same object, e. g. an offside wing is often not visible when shielded by the fuselage;

(e) reflection - other aircraft/vehicle (s) and large structures such as hangars may reflect some energy away from the radar antenna, e. g. a smooth aircraft fuselage at angles other than a right angle to the radar;

(f) rough surfaces or long grass - vehicle detectability is reduced on rough ground, wet or long grass;

(g) radar position elongation - occurs in both range and azimuth, due to radar equipment resolution limitations associated with stronger returns; and

(h) lack of radar position labels and symbols.

SMR cannot be used by ATC to provide heading instructions for taxi guidance. Taxi guidance instructions using SMR should be the same as those applicable for visual control.

The workload and concentration involved in detailed SMR monitoring are significant and can restrict controller traffic handling capacity.

6. 7. 2. 3   Methods of establishing SMR identification

Before providing guidance to an aircraft/vehicle based on SMR-derived position information, positive radar identification is established by use of at least one of the methods specified below:

(a) correlating the position of a visually observed aircraft/vehicle to that displayed by SMR;

(b) correlating an identified SMR position observed from another radar source;

(c) correlating an SMR position complying with an ATC instruction for a specific manoeuvre;

(d) correlating a displayed SMR position to an aircraft or vehicle as reported by radio;

(e) correlating a displayed SMR position to an aircraft or vehicle position;

(f) entering a runway or taxiway intersection;

(g) abeam a building or airfield feature which either shows as a permanent echo on the display, or is marked on the video or grid map; and

(h) on a taxiway or runway, provided that there are no other unidentified vehicles or aircraft on that runway or taxiway segment.

6. 7. 2. 4   Relay of SMR position identification

Positive identification of an SMR-derived aircraft/vehicle position can be relayed by use of the following methods:

(a) direct designation; or

(b) specifying the location of the SMR-derived position by reference to identifiable features displayed on the video or grid map.

The appropriate ATS unit is normally responsible for operating the visual components of the system, including stop bars, taxiway centre line lights and routing designators.

6. 8   Low visibility operations

6. 8. 1   Control of aerodrome surface traffic in low visibility

When there is a requirement for traffic to operate on the maneuvering area in conditions of visibility which prevent the aerodrome control tower from applying visual separation between aircraft, or between aircraft and vehicles, the following procedures can be applied:

(a) At the intersection of taxiways, an aircraft or vehicle on a taxiway cannot to and cannot be permitted by aerodrome controller to hold closer to the other taxiway than the holding position limit defined by a clearance bar, stop bar or taxiway intersection marking according to relevant specifications.

(b) The longitudinal separation on taxiways is as specified for each particular aerodrome by the appropriate ATS authority. This separation takes into account the characteristics of the aids available for surveillance and control of ground traffic, the complexity of the aerodrome layout and the characteristics of the aircraft using the aerodrome.

6. 8. 2   Control of aerodrome traffic for category II/III operation

According to ICAO standards, CAAC shall establish provisions applicable to the start and continuation of precision approach category II/III operations as well as departure operations in RVR conditions less than a value of 550 m.

Low visibility operations are initiated by or through the aerodrome control tower. The aerodrome control tower informs concerned approach control unit when procedures for precision approach category II/III and low visibility operations will be applied and also when such procedures are no longer in force.

The provisions regarding low visibility operations specify the following:

(a) the RVR value (s) at which the low visibility operations procedures are implemented;

(b) the minimum ILS/MLS equipment requirements for category II/III operations;

(c) other facilities and aids required for category II/III operations, including aeronautical ground lights, which is monitored for normal operation;

(d) the criteria for and the circumstances under which downgrading of the ILS/MLS equipment from category II/III operations capability is made;

(e) the requirement to report any relevant equipment failure and degradation, without delay, to the flight crews concerned, the approach control unit, and any other appropriate organization;

(f) applicable spacing between successive approaching aircraft;

(g) action (s) to be taken in the event low visibility operations need to be discontinued, e. g. due to equipment failures; and

(h) special procedures for the control of traffic on the manoeuvring area, including:

1) the runway-holding positions to be used;

2) the minimum distance between an arriving and a departing aircraft to ensure pro-

tection of the sensitive and critical areas;

　　3) procedures to verify that aircraft and vehicles have vacated the runway;

　　4) procedures applicable to the separation of aircraft and vehicles;

　　(i) any other relevant procedures or requirements.

Aerodrome control tower establish a record of vehicles and persons currently on the manoeuvring area and maintain this record during the period of application of these procedures to assist in assuring the safety of operations on that area prior to a period of application of low visibility procedures.

6.9　Parallel or near-parallel operations

Where parallel or near-parallel runways are used for simultaneous operations, the following requirements and procedures apply.

6.9.1　Minimum distance between parallel runways

Where parallel non-instrument runways are intended for simultaneous use, the minimum distance between their centre lines:

— 210 m where the higher code number is 3 or 4;

— 150 m where the higher code number is 2; and

— 120 m where the higher code number is 1.

Where parallel instrument runways are intended for simultaneous use, the minimum distance between their centre lines:

— 1035 m for independent parallel approaches;

— 915 m for dependent parallel approaches;

— 760 m for independent parallel departures;

— 760 m for segregated parallel operations.

except that:

(a) for segregated parallel operations the specified minimum distance:

1) may be decreased by 30 m for each 150 m that the arrival runway is staggered toward the arriving aircraft, to a minimum of 300 m; and

2) increased by 30m for each 150m that the arrival runway is staggered away from the arriving aircraft;

(b) for independent parallel approaches, combinations of minimum distances and associated conditions may be applied when it is determined that such combinations would not adversely affect the safety of aircraft operations.

Wake turbulence categorization of aircraft and wake turbulence separation minima are considered in accordance with the contents in Chapter 5.

6.9.2   Parallel operation for departing aircraft

6.9.2.1   Types of operation

Parallel runways may be used for independent instrument departures as follows:

(a) both runways are used exclusively for departures (independent departures);

(b) one runway is used exclusively for departures while the other runway is used for a mixture of arrivals and departures (semi-mixed operation); and

(c) both runways are used for mixed arrivals and departures (mixed operation).

6.9.2.2   Requirement and procedures for independent parallel departures

Independent IFR departures may be conducted from parallel runways provided:

(a) the runway centre lines are spaced by the distance abovementioned;

(b) the departure tracks diverge by at least 15 degrees immediately after take-off;

(c) suitable surveillance radar capable of identification of the aircraft within 2 km (1.0 NM) from the end of the runway is available; and

(d) ATS operational procedures ensure that the required track divergence is achieved.

6.9.3   Parallel operation for arriving aircraft

6.9.3.1   Types of operations

Parallel runways may be used for simultaneous instrument operations for:

(a) independent parallel approaches; or

(b) dependent parallel approaches; or

(c) segregated parallel operations.

Whenever parallel approaches are carried out, separate radar controllers should be responsible for the sequencing and spacing of arriving aircraft to each runway.

6.9.3.2   Requirements and procedures for independent parallel approaches

Independent parallel approaches may be conducted to parallel runways provided that:

(a) the runway centre lines are spaced by the distance abovementioned; and

1) where runway centre lines are spaced by less than 1310 m but not less than

1 035m, suitable secondary surveillance radar (SSR) equipment, with a minimum azimuth accuracy of 0. 06 degrees (one sigma), an update period of 2. 5 seconds or less and a high resolution display providing position prediction and deviation alert, is available; or

2) where runway centre lines are spaced by less than 1525m but not less than 1 310 m, SSR equipment with performance specifications other than the foregoing may be applied, provided they are equal to or better than those stated under the following below, and when it is determined that the safety of aircraft operation would not be adversely affected; or

3) where runway centre lines are spaced by 1525m or more, suitable surveillance radar with a minimum azimuth accuracy of 0. 3 degrees (one sigma) or better and update period of 5 seconds or less is available;

(b) instrument landing system (ILS) and/or microwave landing system (MLS) approaches are being conducted on both runways;

(c) the missed approach track for one approach diverges by at least 30 degrees from the missed approach track of the adjacent approach;

(d) obstacle survey and evaluation is completed, as appropriate, for the areas adjacent to the final approach segments;

(e) aircraft are advised of the runway identification and ILS localizer or MLS frequency as early as possible;

(f) radar vectoring is used to intercept the ILS localizer course or the MLS final approach track;

(g) no-transgression zone (NTZ) at least 610m (2000ft) wide is established equidistant between extended runway centre lines and is depicted on the radar display;

(h) separate radar controllers monitor the approaches to each runway and ensure that when the 300 m (1000 ft) vertical separation is reduced:

1) aircraft do not penetrate the depicted NTZ; and

2) the applicable minimum longitudinal separation between aircraft on the same ILS localizer course or MLS final approach track is maintained; and

(i) if no dedicated radio channels are available for the radar controllers to control the aircraft until landing:

1) transfer of communication of aircraft to the respective aerodrome controller's frequency is effected before the higher of two aircraft on adjacent final approach tracks intercepts the ILS glide path or the specified MLS elevation angle; and

2) radar controllers monitoring the approaches to each runway are provided with the capability to override transmissions of aerodrome control on the respective radio channels for each arrival flow.

As early as practicable after an aircraft has established communication with approach control, the aircraft is advised that independent parallel approaches are in force. This information may be provided through the ATIS broadcasts.

When vectoring to intercept the ILS localizer course or MLS final approach track, the final vector enable the aircraft to intercept the ILS localizer course or MLS final approach track at an angle not greater than 30 degrees and to provide at least 2 km (1.0 NM) straight and level flight prior to ILS localizer course or MLS final approach track intercept. The vector also enable the aircraft to be established on the ILS localizer course or MLS final approach track in level flight for at least 3.7 km (2.0 NM) prior to intercepting the ILS glide path or specified MLS elevation angle.

A minimum of 300 m (1 000 ft) vertical separation or, subject to radar system and radar display capabilities, a minimum of 5.6 km (3.0NM) radar separation is provided until aircraft are established:

(a) inbound on the ILS localizer course and/or MLS final approach track; and

(b) within the normal operating zone (NOZ).

Subject to radar and display system capabilities, a minimum of 5.6 km (3.0NM) radar separation is provided between aircraft on the same ILS localizer course or MLS final approach track unless increased longitudinal separation is required due to wake turbulence or for other reasons.

An aircraft established on an ILS localizer course or MLS final approach track is separated from another aircraft established on an adjacent parallel ILS localizer course or MLS final approach track provided neither aircraft penetrates the NTZ as depicted on the radar display.

When assigning the final heading to intercept the ILS localizer course or MLS final approach track, the runway is confirmed, and the aircraft is advised of:

(a) its position relative to a fix on the ILS localizer course or MLS final approach track;

(b) the altitude to be maintained until established on the ILS localizer course or MLS final approach track to the ILS glide path or specified MLS elevation angle intercept point; and

(c) if required, clearance for the appropriate ILS or MLS approach.

All approaches regardless of weather conditions are radar-monitored. Control instructions and information necessary to ensure separation between aircraft and to ensure aircraft do not enter the NTZ shall be issued.

The primary responsibility for navigation on the ILS localizer course and/or MLS final approach track rests with the pilot. Control instructions and information are therefore issued only to ensure separation between aircraft and to ensure that aircraft do not penetrate the NTZ.

For the purpose of ensuring an aircraft does not penetrate the NTZ, the aircraft is considered to be the centre of its radar position symbol. However, the edges of the radar position symbols representing aircraft executing parallel approaches are not allowed to touch.

When an aircraft is observed to overshoot the turn-on or to continue on a track which will penetrate the NTZ, the aircraft need to be instructed to return immediately to the correct track.

When an aircraft is observed penetrating the NTZ, the aircraft on the adjacent ILS localizer course or MLS final approach track is instructed to immediately climb and turn to the assigned altitude/height and heading in order to avoid the deviating aircraft. Where parallel approach obstacle assessment surfaces criteria are applied for the obstacle assessment, aerodrome controller cannot issue the heading instruction to the aircraft below 120m (400ft) above the runway threshold elevation, and the heading instruction cannot exceed 45 degrees track difference with the ILS localizer course or MLS final approach track.

Radar monitoring cannot be terminated until:

(a) visual separation is applied, provided procedures ensure that both radar controllers are advised whenever visual separation is applied;

(b) the aircraft has landed, or in case of a missed approach, is at least 2 km (1.0 NM) beyond the departure end of the runway and adequate separation with any other traffic is established. There is no requirement to advise the aircraft that radar monitoring is terminated.

6.9.3.3 Suspension of independent parallel approaches to closely-spaced parallel runways

Independent parallel approaches to parallel runways spaced by less than 1525 m between their centre lines is suspended under certain weather conditions, as prescribed by state ATS authority including windshear, turbulence, downdrafts, crosswind and severe weather such as thunderstorms, which might otherwise increase ILS localizer course and/or MLS final approach track deviations to the extent that safety may be impaired. The increase in final approach track deviations would additionally result in an unacceptable level of deviation alerts being generated.

6.9.3.4 Requirements and procedures for dependent parallel approaches

Dependent parallel approaches may be conducted to parallel runways provided:

(a) the runway centre lines are spaced by the distance specified abovementioned;

(b) the aircraft are radar vectored to intercept the final approach track;

(c) suitable surveillance radar with a minimum azimuth accuracy of 0.3 degrees (one sigma) and update period of 5 seconds or less is available;

(d) ILS and/or MLS approaches are being conducted on both runways;

(e) aircraft are advised that approaches are in use to both runways (this information may be provided through the ATIS);

(f) the missed approach track for one approach diverges by at least 30 degrees from the missed approach track of the adjacent approach; and

(g) approach control has a frequency override capability to aerodrome control.

A minimum of 300 m (1000 ft) vertical separation or a minimum of 5.6 km (3.0 NM) radar separation is provided between aircraft during turn-on to parallel ILS localizer courses and/or MLS final approach tracks.

The minimum radar separation to be provided between aircraft established on the ILS localizer course and/ or MLS final approach track shall be:

(a) 5.6 km (3.0NM) between aircraft on the same ILS localizer course or MLS final approach track unless increased longitudinal separation is required due to wake turbulence; and

(b) 3.7km (2.0 NM) between successive aircraft on adjacent ILS localizer courses or MLS final approach tracks.

6.9.3.5   Requirements and procedures for segregated parallel operations

Segregated parallel operations may be conducted on parallel runways provided:

(a) the runway centre lines are spaced by the distance abovementioned; and

(b) the nominal departure track diverges immediately after take-off by at least 30 degrees from the missed approach track of the adjacent approach (see Fig 6.3) .

The minimum distance between parallel runway centre lines for segregated parallel operations may be decreased by 30 m for each 150 m that the arrival runway is staggered toward the arriving aircraft, to a minimum of 300 m (see Fig 6.4) and should be increased by 30m for each 150 m that the arrival runway is staggered away from the arriving aircraft (see Fig 6.5) .

The following types of approaches may be conducted in segregated parallel operations provided suitable surveillance radar and the appropriate ground facilities conform to the standard necessary for the specific type of approach:

(a) ILS and/or MLS precision approach;

(b) surveillance radar approach (SRA) or precision approach radar (PAR) approach; and

(c) visual approach.

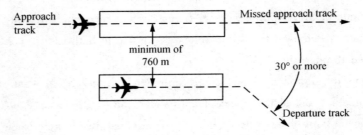

Fig 6. 3   Segregated parallel operations

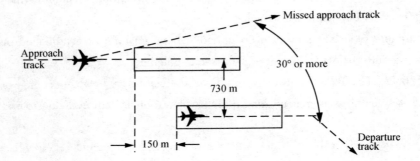

Note—In the event of a missed approach by a heavy jet aircraft, wake turbulence separation should be applied or, alternatively, measures taken to ensure that the heavy jet aircraft does not overtake an aircraft departing from the adjacent parallelrunway.

Fig 6. 4   Segregated parallel operations where runways are staggered

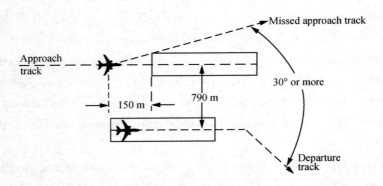

Fig 6. 5   Segregated parallel operations where runways are staggered

6. 10   Aeronautical ground lighting operation

Aeronautical ground lights are operated, except as following:

(a) continuously during the hours of darkness or during the time the centre of the sun's disc is more than 6 degrees below the horizon, whichever requires the longer period of operation, unless otherwise provided hereafter or otherwise required for the control of air traffic;

(b) at any other time when their use, based on weather conditions, is considered desirable for the safety of air traffic.

Lights on and in the vicinity of aerodromes that are not intended for en-route navigation purposes may be turned off, subject to further provisions hereafter, if no likelihood of either regular or emergency operation exists, provided that they can be again brought into operation at least one hour before the expected arrival of an aircraft.

At aerodromes equipped with lights of variable intensity a table of intensity settings, based on conditions of visibility and ambient light, are provided for the guidance of aerodrome controllers in effecting adjustment of these lights to suit the prevailing conditions. When so requested by an aircraft, further adjustment of the intensity is made whenever possible.

6. 10. 1   Approach lighting operations

Approach lighting hereunder includes such lights as simple approach lighting systems, precision approach lighting systems, visual approach slope indicator systems, circling guidance lights, approach light beacons and runway alignment indicators.

In addition to the above, approach lighting is also operated:

(a) by day when requested by an approaching aircraft;

(b) when the associated runway lighting is operated.

The lights of a visual approach slope indicator system are operated during the hours of daylight as well as of darkness and irrespective of the visibility conditions when the associated runway is being used.

6. 10. 2   Runway lighting operations

Runway lighting includes such lights as edge, threshold, centre line, end, touchdown zone and wing bar lights.

Runway lighting need not be operated if that runway is not in use for landing, take-off or taxiing purposes, unless required for runway inspections or maintenance.

If runway lighting is not operated continuously, lighting following a take-off is provided as specified below:

(a) at aerodromes where air traffic control service is provided and where lights are centrally controlled, the lights of one runway remain lighted after take-off as long as is considered necessary for the return of the aircraft due to an emergency occurring during or immediately after take-off;

(b) at aerodromes without air traffic control service or without centrally controlled lights, the lights of one runway remain lighted until such time as would normally be required to reactivate the lights in the likelihood of the departing aircraft returning for an emergency landing, and in any case not less than fifteen minutes after take-off.

Where obstacle lighting is operated simultaneously with runway lighting, particular care should be taken to ensure that it is not turned off until no longer required by the aircraft.

6. 10. 3　Stopway lighting operations

Stopway lights need to be operated whenever the associated runway lights are operated.

6. 10. 4　Taxiway lighting operations

Taxiway lighting includes such lights as edge lights, centre line lights, stop bars and clearance bars.

Where required to provide taxi guidance, taxiway lighting need to be turned on in such order that a continuous indication of the taxi path is presented to taxiing aircraft. Taxiway lighting or any portion thereof may be turned off when no longer needed.

6. 10. 5　Stop bars operations

Stop bars are located across taxiways at the point where it is desired that traffic stop, and consist of lights, showing red, spaced across the taxiway. Stop bars are switched on to indicate that all traffic stop and switched off to indicate that traffic may proceed.

6. 10. 6　Obstacle lighting operations

Obstacle lighting includes such lights as obstacle and unserviceable lights and hazard beacons.

Obstacle lighting associated with the approach or departure from a runway or channel, where the obstacle does not project through the inner horizontal surface, they may be turned off and on simultaneously with the runway or channel lights. Unserviceable lights may not be turned off as permitted while the aerodrome is open.

6. 10. 7   Monitoring of visual aids

Aerodrome controllers make use of automatic monitoring facilities, when provided, to ascertain whether the lighting is in good order and functioning according to selection.

In the absence of an automatic monitoring system or to supplement such a system, the aerodrome controller can visually observe such lighting as can be seen from the aerodrome control tower and use information from other sources such as visual inspections or reports from aircraft to maintain awareness of the operational status of the visual aids.

On receipt of information indicating a lighting fault, the aerodrome controller need to take such action as is warranted to safeguard any affected aircraft or vehicles, and initiate action to have the fault rectified.

6. 11   Runway incursion prevention

"Runway incursions" are defined in a number of different ways within the aviation industry. The definition by ICAO is any occurrence at an aerodrome involving the incorrect presence of an aircraft, vehicle, person on the protected area of a surface designated for the landing and take-off of aircraft. Such incidences are incredibly costly for airlines and, more importantly, have a high risk of causing fatalities. According to comparative data analysis results, the number of nationwide runway incursions has significantly increased as traffic volume increases.

Pilots, controllers, and drivers can all be involved in runway incursions. In October 2002, ICAO presented a plan to address runway incursions. This plan was based on a system safety approach that would identify actual and potential hazards, provide remedial action and the monitoring and assessment of hazards in the following areas:

(a) Radiotelephony phraseology

(b) Language proficiency

(c) ATC procedures

(d) Performance requirements for equipment

(e) Aerodrome lighting and markings

(f) Aerodrome charts

(g) Operational aspects

(h) Situational awareness

(i) Human factors

6. 12　Aerodrome alerting service

Generally, rescue coordination centre of aerodrome authority is the unit responsible for promoting efficient organization of search and rescue services and for coordinating the conduct of search and rescue operations while aircraft encountering emergency situation in vicinity of the aerodrome. Aerodrome control tower is responsible for provision of alerting service to notify appropriate organizations regarding aircraft in need of search and rescue aid, and assist such organizations as required.

Aerodrome control tower provides alerting as well as fire fighting services whenever:

(a) an aircraft accident has occurred on or in the vicinity of the aerodrome;

(b) information is received that the safety of an aircraft which is or will come under the jurisdiction of the aerodrome controller may have or has been impaired; or

(c) requested by pilot;

(d) when otherwise deemed necessary or desirable.

Exercise

I. Explain the following definitions.

(1) traffic circuit (2) segregated parallel runway (3) NOZ

(4) NTZ　　　　(5) control zone　　　　(6) visual approach

(7) Low visibility operation

II. Answer the following questions.

(1) Describe the main functions of an aerodrome control tower.

(2) Please give the main principle for aerodrome controller to select a "runway-in-use".

(3) According to ICAO standards, what's the corresponding Airspace Class with terminal (approach) control area and aerodrome control tower area?

（4）Describe visual approach application procedure.

（5）What's the minimum distance between parallel runways centre line for instrument operations?

Ⅲ. **Draw out the tracks of the aircraft entering traffic circuit by dash line on the figure based on the given ATC instructions.**

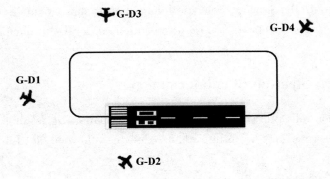

（1）CTL：G-D1，straight in approach runway 09，report on final.

（2）CTL：G-D2，descend to 300 meters，join left hand traffic circuit runway 09，report on downwind.

（3）CTL：G-D3，make a short approach runway 09，report turning base.

（4）CTL：G-D4，make a right-hand orbit at present position，join left hand traffic circuit runway 09，report on downwind.

Chapter 7

Aerodrome control service

7.1 Provision of information to aircraft

Aerodrome controller provides relevant information such as weather condition and local traffic information, which related to aircraft operations described in chapter 6 to aircraft as following procedure.

7.1.1 Forward Start-up time information

Aerodrome controller forwards Start-up time information by issuing a Start-up Clearance to pilot requesting Start-up in accordance with the allocated slot time as the following phraseologies:

PIL: [aircraft location] Request Start-up;

PIL: [aircraft location] Request Start-up, Information (ATIS identification);

CTL: Start-up Approved.

When clear pilot to start-up at his/her own discretion:

CTL: Start-up at Own Discretion;

CTL: Expect Departure (time) Start-up at Own Discretion.

When issuing an expected start-up time to pilot requesting start-up:

CTL: Start-up at (time);

CTL: Expect Start-up at (time).

Example 1:

➕ " Pudong Ground, CSN6213, gate 15, request start up, with information Bravo. "

☆ " CSN6213, start up approved. "

Example 2:

➕ " Pudong Ground, CSN6213, gate 11, request start up. "

☆ "CSN6213, start up at 30, QNH 1017. "

Example 3:

✈ "Pudong Ground, Air China 1256, gate 31, request start up. "

☆ "Air China 1256, expect departure at 49, start up at own discretion. "

7.1.2　Forward essential information

Generally, aerodrome controller provides the essential information which includes aerodrome condition, local traffic, meteorological and other relevant information to a departing aircraft prior to taxiing or on requesting by the pilot.

For arriving aircraft, the above essential information is provided prior to the aircraft joining traffic circuit or commencing its approach to land by approach controller if exist. However, some of information is only be informed to the aircrafts concerned.

When issuing the essential meteorological information:

CTL: Runway in use (number); Surface Wind (number) Degrees, (number) meters Per Second (or Knots); Visibility (distance) (units), or RVR (or Runway Visual Range) [RUNWAY (number)] (first position) (distance) (units), (second position) (distance) (units), (third position) (distance) (units); Present Weather (details); Cloud (amount, [(type)] and height of base) (units) (or Sky clear); Temperature [Minus] (number) and/or Dew-Point [Minus] (number); QNH or QFE is (number);

CTL: Tower Observes (weather information);

CTL: Pilot Reports (weather information) .

Example 4:

✈ "Pudong Ground, Air China 824, request weather information for Shanghai. "

☆ "Air China 824, Shanghai weather at 1430, ceiling 170 meters overcast, imbedded thunderstorms, ceiling 500 meters, visibility 5 kilometers, temperature minus 2, wind 310 degrees at 4meters per second, QNH 1021. "

When changing from IFR flight to VFR flight while it is likely that flight in VMC cannot be maintained:

CTL: Instrument Meteorological Conditions Reported (or forecast) in the Vicinity of (location) .

When issuing the essential aerodrome conditions information:

CTL: [(location)] Runway Surface Condition Runway (number) (condition); or

[(location)] Runway Surface Condition Runway (number) Not Current (When surface condition is not available);

CTL: Landing Surface (condition);

CTL: Caution Construction Work (location); or Caution (specify reasons) Right (or Left), (or Both Sides) of Runway [number]; or Caution Work in Progress (or Obstruction) (position and any necessary advice);

CTL: Runway Report at (observation time) Runway (number) (type of precipitant) up to (depth of deposit) Millimeters. Braking Action Good (or Medium to Good, or Medium, or Medium to Poor, or Poor or Unreliable) [and/or Braking Coefficient (equipment and number)];

CTL: Braking Action Reported by (aircraft type) at (time) Good (or Medium, or Poor);

CTL: Breaking Action [(location)] (measuring equipment used), Runway (number), Temperature [Minus] (number), Was (reading) at (time);

CTL: Runway (or Taxiway) (number) Wet [or Damp, Water Patches, Flooded (depth), or Snow Removed (length and width as applicable), or Treated, or Covered With Patches of Dry Snow (or Wet Snow, or Compacted Snow, or Slush, or Frozen Slush, or Ice, or Ice Underneath, or Ice and Snow, or Snowdrifts, or Frozen Ruts and Ridges)];

Example 5:

✛ "Shenyang Tower, Asiana 784, short final, request runway surface conditions. "

☆ "Asiana 784, Runway 17 wet, covered with patches of snow and ice. A landing B737 reported frozen ruts and ridges on runway. "

Example 6:

✛ "Air China 879, caution, obstruction west side of taxiway C. Move to left side of taxiway for clearance. "

☆ "Obstruction in sight, Move to left side of taxiway C, Air China 879. "

When issuing the essential local traffic information:

CTL: Traffic (information);

CTL: No Reported Traffic;

PIL: Looking Out;

PIL: Traffic in Sight;

PIL: Negative Contact [reasons];

CTL: [Additional] Traffic (direction) Bound (type of aircraft) (level) Estimated （or Over） （significant point） at （time）；

　　CTL：Traffic is （classification） Unmanned Free Balloon （s） was ［or Estimated］ Over （place） at （time） Reported （level （s）） ［or Level Unknown］ Moving （direction） （other pertinent information, if any） .

Example 7：

　　☆ "United 135, Pudong Tower, extend downwind, traffic B757 on short final, report traffic in sight. "

　　✗ "Pudong Tower, United 135, traffic in sight. "

When issuing operational abnormal status of visual and non-visual aids：

　　CTL： （specify visual or non-visual aid） Runway （number） （description of deficiency）；

　　CTL： （type） Lightning （unserviceable）；

　　CTL： MLS/ILS Category （category） （serviceability state）；

　　CTL： Taxiway Lighting （description of deficiency）；

　　CTL： （type of visual approach slope indicator） Runway （number） （description of deficiency） .

Example 8：

　　☆ "United 56 heavy, Pudong Tower, ILS category I minimums in effect. ILS category II is unavailable due to inoperative centre line lighting. "

　　✗ "ILS category I minimums, United 56 heavy. "

Example 9：

　　☆ "China Southern 6019, Pudong Tower, approach lighting inoperative. "

　　✗ "Approach lighting inoperative, China Southern 6019. "

SUMMARY：

　　Essential aerodrome conditions information includes：

- Construction or maintenance work；
- Rough or broken surfaces on a runway, taxiway or apron；
- Snow, slush or ice on runway, taxiway or apron；
- Water on a runway, taxiway or apron；
- Snow banks or drifts adjacent to a runway, taxiway or an apron；
- Temporary hazards, including birds on the ground or in the air；
- Failure or irregular operation of part/all of the aerodrome lighting/navigation system；
- Any other pertinent information.

ATIS broadcast

At some busy aerodromes, essential information may be broadcasted on ATIS message alternated by English and Chinese. The contents of ATIS generally include the following items:

(Aerodrome identification) Information (Alphabet). (Time), Wind (number) degrees at (number) (unit), Visibility (number) meters. Ceiling (number), Description of the ceiling. Temperature (number), Dew point (number), QNH or QFE (number), (Type of Approach) Runway in use (number), 〔Departing Runway (number)〕. Hazardous Weather Information for (geographical area) (details). Advise on initial contact you have (Alphabet).

Example 10:

Pudong airport, information Hotel 1500 UTC. Landing and departing runway 17. ILS runway 17 approach in use. Temperature 18, QNH 1031, Wind 230 degrees at 4 meters / second, gusting to 8 meters / second. Visibility 4000 meters, Lightning reported in vicinity of the airport. Clouds 400 meters scattered, measured ceiling 700 meters broken. Advise on initial contact you have information Hotel.

7.1.3 Other information provision

Aerodrome controller need to transmit the following information related to aircraft operation in the vicinity of aerodrome to pilot without delay:

When pilot requested visual inspection of landing gear:

CTL: Landing Gear Appears Down;

CTL: Right (or Left, or Nose) Wheel Appears up (or down);

CTL: Wheels Appear up;

CTL: Right (or Left, or Nose) Wheel Does not Appear up (or down).

Example 11:

✗ "Pudong Tower, United 566, request low pass, unsafe left gear indication."

☆ "United 566, cleared low pass Runway 17, not below 100 meters, report final."

✗ "Runway17 not below 100 meters, report final, United 566."

✗ "Pudong Tower, United 566, short final."

☆ "United 566, roger, left gear appears up."

When issuing wake turbulence warning:

CTL: Caution Wake Turbulence 〔From Arriving (or Departing) (type of aircraft)〕

[additional information as required] .

When issuing jet blast on apron or taxiway warning:

CTL: Caution Jet Blast.

When issuing propeller-driven aircraft slipstream:

CTL: Caution Slipstream.

Example 12:

☆ "China Southern 6005, cleared for takeoff, caution wake turbulence. "

✛ "Cleared for take off, China Southern 6005. "

Example 13:

★ "Ground, truck 9, request to proceed to maintenance base. "

☆ "Truck 9, give way to the United Boeing 767 on your right then proceed to maintenance base, caution jet blast. "

7.2　Control of aerodrome traffic

As the view of the pilot from the flight deck is normally restricted, aerodrome controller needs to ensure that instructions and information which require pilot to employ visual detection, recognition and observation are issued in a clear, concise and complete way.

7.2.1　Designated positions in aerodrome control

The following positions in the traffic and taxi circuits are the positions where aerodrome controller issues clearances to pilot. Controller is able to watch closely as they approach these positions so that proper clearances can be issued without delay. Where practicable, all clearances can be issued without waiting for pilot request (Fig 7.1) .

Position 1. For departing flight, pilot initiates call for taxi instructions. Aerodrome controller normally issues runway-in-use information and taxi clearances here.

Position 2. Departing aircraft holds at this position when there is conflicting traffic until cleared by controller.

Position 3. Aerodrome controller generally issues take-off clearance here, when it's not practicable at position 2.

Position 4. Clearance to land is issued here as practicable.

Position 5. Clearance to taxi to apron is issued here.

Position 6. Parking information issued here.

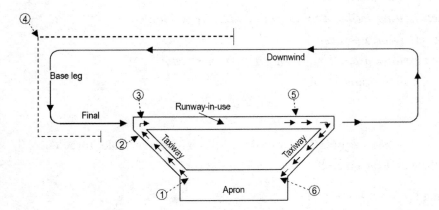

Fig 7. 1   Designated positions of aircraft from the controller viewpoint

Arriving aircraft executing an instrument approach procedure normally enters the traffic circuit on final except when visual maneuvering to the landing runway is required.

7. 2. 2   Control of the taxing aircraft

7. 2. 2. 1   Taxi and pushback clearance and instructions

Taxi clearances issued by aerodrome controller contain concise instructions and adequate information so as to assist pilot to follow the correct taxi routes, to avoid collision with other aircraft or objects and to minimize the potential for the aircraft inadvertently entering an active runway. When taxi clearance contains taxi limit beyond runway, aerodrome controller issues the clearance in an explicit way to cross or hold short of that runway. Prior to issuing taxi clearance, controller determines the position that the concerned aircraft parked.

Standard taxi routes used at an aerodrome is published in the national AIP by state ATS authority whenever practicable. They are identified by designators and be used in taxi clearances. Where there is no standard taxi routes published, aerodrome controller issues taxi route described by use of taxiway and runway designators. Other relevant information, such as aircraft to follow or give way to, is also provided to a taxiing aircraft.

When pilot request pushback clearance or instructions:

PIL: [aircraft location] Request Pushback;

CTL: Pushback Approved;

CTL: Stand by;

CTL: Pushback at Own Discretion;

CTL: Expect (number) Minutes Delay Due (reason) .

Example 1:

✈ "Pudong Ground, CSN6501, gate 31, request push back. "

☆ "CSN6501, standby. "

☆ "CSN6501, push back approved. "

When pilot request taxi clearance or instructions:

PIL: [aircraft type] [wake turbulence category if "heavy"] [aircraft location] Request Taxi [intentions];

PIL: [aircraft type] [wake turbulence category if "heavy"] [aircraft location] (flight rules) to (aerodrome of destination) Request Taxi [intentions];

PIL: [aircraft type] [wake turbulence category if "heavy"] Request Detailed Taxi Instructions;

PIL: [(aircraft location)] Request Taxi to (destination on aerodrome) .

When controller issuing taxi clearance or instructions:

CTL: Taxi to Holding Position [number] [Runway (number)] [Time (time)];

CTL: Taxi to Holding Position [(number)] [Runway (number)] Via (specific route to be followed) [Time (time)] [Hold Short of Runway (number)];

CTL: Taxi to Holding Position [(number)] (followed by aerodrome information as applicable) [Time (time)];

CTL: Take (or Turn) First (or Second) Left (or Right);

CTL: Taxi Via (identification of taxiway);

CTL: Taxi Via Runway (number);

CTL: Taxi to Terminal (or other location, e. g. General Aviation Area)　[Stand (number)];

CTL: Taxi Straight Ahead;

CTL: Taxi with Caution;

CTL: Give Way to (description and position of other aircraft);

CTL: Taxi into Holding Bay;

CTL: Follow (description of other aircraft or vehicle);

CTL: Expedite Taxi [(reason)];

CTL: [Caution] Taxi Slower [reason];

PIL: Slowing down.

Example 2:

✈ "Pudong ground, United 428 heavy, Stand 11, request taxi. "

☆ "United428 heavy, taxi to holding point, runway35, wind 085 degrees at 7 meters per second, QNH 1023. "

✚ "Roger, taxiing to runway35, QNH 1023. Delta 428. "

Example 3:

✚ "Pudong ground, United511, bridge 10, request taxi. "

☆ " United 511, taxi via K5, T5 to runway 36. "

✚ "Roger, taxiing to runway36, United 511. "

☆ "United 511, hold short of T5, give way to China Southern B757 proceeding west to parking. "

When issuing back-track clearance or instructions:

PIL: Request Backtrack;

CTL: Backtrack Approved;

CTL: Backtrack Runway (number) .

Example 4:

✚ "Pudong tower, China Southern 6012, request backtrack. "

☆ "China Southern 6012, backtrack Runway 17 vacating Runway. "

When issuing tow clearance or instructions:

PIL: Request Tow [company name] (aircraft type) from (location) to (location);

CTL: Tow Approved Via (specific routing to be followed);

CTL: Hold Position;

CTL: Stand by.

Example 5:

✶ "Ground, tug 6, request tow United Boeing 757 from maintenance hangar 3 to gate 25. "

☆ "Tug 6, tow approved to gate 25 via west end of apron. "

7.2.2.2 Use of runway-holding positions

Aircraft cannot hold closer to a runway-in-use beyond runway-holding position when it taxi through the runway.

Aerodrome controller never permits pilot to line up and hold on the approach end of a runway-in-use whenever another aircraft is affecting landing, until the landing aircraft has passed the point of intended holding (see Fig 7.2).

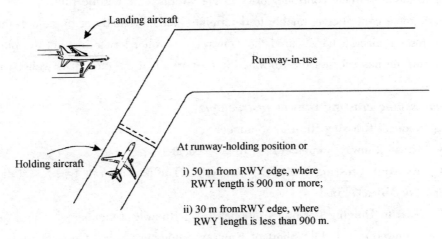

Fig 7.2   Method of holding aircraft

When issuing holding instructions:

CTL: Hold (direction) of (position, runway number, etc.);

CTL: Hold Position;

CTL: Hold (distance) from (position);

CTL: Hold Short of (position);

PIL: Holding;

PIL: Holding Short.

Example 6:

✦ "Pudong Ground, United 428 heavy, stand 11, request taxi."

☆ "United 428 heavy, hold on taxiway E for ambulance proceeding to runway 17."

Example 7:

"Pudong Tower, Japan Air 47, ready for takeoff."

☆ "Japan Air 47, wait, hold short of runway."

7.2.2.3   Taxiing instructions on a runway-in-use

For expediting air traffic, aerodrome controller may permit aircraft taxi on the runway-in-use, provided no delay or risk to other aircraft. Where control of taxiing aircraft is provided by a ground controller while the control of runway operations by an aerodrome controller, the use of a runway is coordinated between ground controller and approved by aerodrome controller, communication with the aircraft concerned is handover from the ground controller to the aerodrome controller prior to the aircraft entering the runway.

If aerodrome controller is unable to determine, either visually or by radar, that a vacating or crossing aircraft has cleared the runway, controller has to instruct the pilot to report when he/she has vacated the runway. Pilot report when the aircraft is well clear of the runway.

When issuing crossing runway instructions:

PIL: Request Crossing Runway (number);

CTL: Cross Runway (number), [Report Vacated];

CTL: Expedite Crossing Runway (number), Traffic (aircraft type), (distance) Kilometers (or Miles) Final;

CTL: Taxi to Holding Position [number], [Runway (number)], Via (specific route to be followed), [Hold Short of Runway (number)] or [Cross Runway (number)].

Example 8:

✦ "Pudong Ground, United 65 heavy, request cross runway 35."

☆ "United 65 heavy, expedite crossing runway 35, traffic B737, five kilometers final."

Runway vacating instructions:

CTL: Vacate Runway;

PIL: Runway Vacated.

Example 9:

✦ "Pudong Ground, Air China952 heavy, request cross runway 17."

☆ "Air China952 heavy, expedite crossing runway 17, report vacate runway."

✦ "Air China952, runway vacated."

7.2.2.4 Helicopter taxing instructions

When a wheeled helicopter or vertical take-off and landing (VTOL) aircraft to taxi on the paved surface, Aerodrome controller may clear the pilot of a helicopter when it is requested or necessary for the helicopter to proceed at a slow speed above the surface [normally below 37 km/h (20 kt) and in ground effect].

Controller has to avoid instructions which require small aircraft or helicopters to taxi in close proximity to taxiing helicopters and consider the effect of turbulence from taxiing helicopters on arriving and departing light aircraft.

　　Controller cannot instruct single-pilot of helicopter frequency change when he/she is hovering or air-taxiing. Whenever possible, control instructions from the next ATS unit need to be relayed as necessary until the pilot is able to change frequency.

　　Ground taxiing uses less fuel than air-taxiing and minimizes air turbulence. However, under certain conditions, such as rough, soft or uneven terrain, it may become necessary to air-taxi for safety considerations. Helicopters with articulating rotors (usually designs with three or more main rotor blades) are subject to "ground resonance" and may, on rare occasions, suddenly lift off the ground to avoid severe damage or destruction.

　　Air-taxiing consumes fuel at a high burn rate, and helicopter downwash turbulence (produced in ground effect) increases significantly with larger and heavier helicopters.

　　When a helicopter request air-taxi clearance or instructions:

　　PIL: Request Air-Taxiing from (or Via) to (location or routing as appropriate);

　　CTL: Air-Taxi to (or Via) (location or routing as appropriate) [Caution (dust, blowing snow, loose debris, taxiing light aircraft, personnel, etc.)];

　　CTL: Air-Taxi Via (direct, as requested, or specified route) to (location, heliport, operating or movement area, active or inactive runway), Avoid (aircraft or vehicles or personnel).

Example 10:

　　🚁 "Pudong Ground, G-CD, request air taxi from apron to maintenance area."
　　☆ "G-CD, air taxi to maintenance area via taxiway C, avoid the parking B757."
　　🚁 "Air taxi to maintenance area via taxiway C, G-CD."

7.2.3　Control of persons and vehicles at aerodrome

　　The movement of persons or vehicles including towed aircraft on the manoeuvring area of an aerodrome is also controlled as necessary to avoid hazard to them or to aircraft landing, taxiing or taking off.

　　In conditions where low visibility procedures are in operation:

　　(a) persons and vehicles operating on the manoeuvring area of an aerodrome are restricted to the essential minimum, and particular concerns will be given to the requirements to protect the ILS/MLS sensitive area (s) when Category II or Category III precision instrument operations are in progress;

　　(b) The minimum separation between vehicles and taxiing aircraft are prescribed by the state ATS authority taking into account the aids available;

(c) when mixed ILS and MLS Category II or Category III precision instrument operations are taking place to the same runway continuously, more restrictive ILS or MLS critical and sensitive areas are protected.

Emergency vehicles proceeding to the assistance of an aircraft in distress are handled with priority over all other surface movement traffic, and vehicles on the manoeuvring area are required to comply with the following rules:

(a) vehicles and vehicles towing aircraft give way to aircraft which are landing, taking off or taxiing;

(b) vehicles give way to other vehicles towing aircraft;

(c) vehicles give way to other vehicles in accordance with controller instructions;

(d) Besides the above items, vehicles and vehicles towing aircraft need to comply with instructions issued by the aerodrome controller.

The movement of pedestrians or vehicles on the manoeuvring area is also subject to authorization from aerodrome controller. Persons, including drivers of all vehicles, are required to obtain authorization from controller before entry to the manoeuvring area. Notwithstanding such an authorization, entry to a runway or runway strip or change in the operation is subject to a further specific authorization by aerodrome controller.

All vehicles and pedestrians give way to aircraft which are landing, taxiing or taking off, except that emergency vehicles proceeding to the assistance of an aircraft in distress have priority over all other surface movement traffic. In the latter case, all movement of surface traffic, to the extent practicable, is halted until it is determined that the progress of the emergency vehicles will not be impeded.

When an aircraft is landing or taking off, vehicles are not permitted to hold closer to the runway-in-use than:

(a) at a taxiway/runway intersection — at a runway holding position; and

(b) at a location other than a taxiway/runway intersection— at a distance equal to the separation distance of the runway-holding position.

At controlled aerodromes all vehicles employed on the maneuvering area is able to maintain two-way radiocommunication with the aerodrome control tower, except when the vehicle is only occasionally used on the maneuvering area and is:

(a) accompanied by a vehicle with the required communications capability, or

(b) employed in accordance with a pre-arranged plan established with the controller.

When communications by a system of visual signals is deemed to be adequate, or in the case of radiocommunication failure, the signals given hereunder have the following meaning:

Light signal from aerodrome control	**Meaning**
Green flashes	Permission to cross landing area or to move onto taxiway
Steady red Stop	Red flashes Move off the landing area or taxiway and watch out for aircraft
White flashes	Vacate manoeuvring area in accordance with local instructions

In emergency conditions or the signals above are not observed, the signal given hereunder used for runways or taxiways equipped with a lighting system with the meaning indicated therein.

Light signal	**Meaning**
Flashing runway or taxiway lights	Vacate the runway and observe the tower or light signal

The phraseologies between aerodrome controller and vehicles are similar with those between towers and aircraft except replacing "Proceed" instead of "Taxi".

7.3   Control of departing aircraft

7.3.1   ATC clearance

The issuance of ATC clearances by air traffic controller constitutes authority for an aircraft to proceed only in so far as known air traffic is concerned. ATC clearances do not constitute authority to violate any applicable regulations for promoting the safety of flight operations or for any other purpose; neither do clearances relieve a pilot-in-command of any responsibility whatsoever in connection with a possible violation of applicable rules and regulations. ATC units issue ATC clearances as are necessary to prevent collisions and to expedite and maintain an orderly flow of air traffic and must be issued early enough to ensure that they are transmitted to the aircraft in sufficient time for it to comply with them. Generally, ATC clearance is issued by the ground or tower controller before aircraft take-off.

If an air traffic control clearance is not suitable to aircraft, pilot may request and, if practicable, obtain an amended clearance.

7.3.1.1   Coordination

ATS units have to establish standardized procedures for standard clearances between the ATC units concerned for departing aircraft.

ACCs forward clearance to approach control or aerodrome control with the least possible delay after receipt of request made by these units, or prior to such request if practicable except where procedures providing for the use of standard departure clearances have been implemented.

Where standard clearances for departing aircraft have been agreed to between concerned ATS units, the aerodrome controller normally issues standard clearance without prior coordination with or approval from the approach control or ACC.

When coordination for release an aircraft:

CTL1: Request Release of (aircraft call sign);

CTL2: (aircraft call sign) Released [at (time)] [conditions/restrictions];

CTL1: Is (aircraft call sign) Released;

CTL2: (aircraft call sign) not Released [Until (time or significant point)];

CTL2: Unable (aircraft call sign) [Traffic is (details)] .

Example 1:

☆ "Approach, Tower, request release of United 453. "

❈ "Tower, United 453 released at 34. "

When coordination for change of clearance:

CTL1: May we Change Clearance of (aircraft call sign) to (details of alteration proposed);

CTL2: Agreed to (alteration of clearance) of (aircraft call sign);

CTL2: Unable (aircraft call sign);

CTL2: Unable (desired route, level, etc.) [for (aircraft call sign)] [due (reason)] (alternative clearance proposed) .

Example 2:

☆ "Approach, Tower, may we change release of United 453 at 28?"

❈ "Tower, unable release United 453 released at 28 due flow control. "

When coordination for approving a request:

CTL1: Approval Request (aircraft call sign) Estimated Departure from (significant point) at (time);

CTL2: (aircraft call sign) Request Approved [(restriction if any)];

CTL2: (aircraft call sign) Unable (alternative instructions) .

When coordination for expedition of clearance:

CTL2: Expedite Clearance (aircraft call sign) Expected Departure from (place) at (time);

CTL2: Expedite Clearance (aircraft call sign)   [Estimated] over (place) at (time) Requests (level or route, etc.) .

Example 3:

☆ "Approach, Tower, Approval Request , Cathy 254, estimated departure at 45. "
❊ "Tower, Cathy 254 request approved. "

7. 3. 1. 2   Standard contents of ATC clearances

Clearances contain positive and concise data and are phrased in a standard manner. Standard clearances for departing aircraft contain the following items:
(a) aircraft identification;
(b) clearance limit, normally destination aerodrome;
(c) designator of the assigned SID, if applicable;
(d) initial level, except when this element is included in the SID description;
(e) allocated SSR code;
(f) any other necessary instructions or information not contained in the SID description, e. g. instructions relating to change of frequency.
When issuing ATC clearance for IFR flight:
CTL: (Aircraft identification) CLEARED (Clearance limit) (Route of flight) (Level assignment) (Additional items)
CTL: (Name of unit) CLEARS (Aircraft identification) (Clearance limit) (Route of flight) (level assignment) (Additional items)

Example 4:

☆ "Air China952, cleared to DaLian via flight planned route, request level change en-route, Squawk 3131. "
✚ "Cleared to DaLian via flight planned route, request level change en-route, Squawk 3131, Air China952. "

7. 3. 1. 3   Read-back of ATC clearances

Pilot has to read back to controller safety-related parts of ATC clearances and instructions which are transmitted by voice. However, the following items are always read back:
(a) ATC route clearances;
(b) clearances and instructions to enter, land on, take off on, hold short of, cross taxi and backtrack on any runway; and

(c) runway-in-use, altimeter settings, SSR codes, level instructions, heading and speed instructions and, whether issued by the controller or contained in automatic terminal information service (ATIS) broadcasts, transition levels.

It is controller's responsibility listen to the read-back for ascertain that the clearance or instruction has been correctly acknowledged by pilot and take immediate action to correct any discrepancies revealed by the read-back.

7.3.2 Take-off clearance

When aerodrome controller ascertains that the separation has existed while the aircraft commences take-off, he/she can issue the take-off clearance to pilot.

When an ATC clearance is required prior to takeoff, aerodrome controller cannot issue take-off clearance until ATC clearance has been transmitted to and acknowledged by pilot. ACC or approach need to forward ATC clearances to aerodrome control tower with the least possible delay after receipt of a request from aerodrome controller or prior to such request if practicable.

Aerodrome controller generally issues take-off clearance when pilot is ready for take-off and at or approaching the departure runway and the traffic situation permits. To reduce the potential for misunderstanding, the take-off clearance issued by controller includes the designator of the departure runway.

To expedite traffic, aerodrome controller may issue an immediate take-off clearance to an aircraft before it enters the runway. On acceptance of such clearance pilot taxis out to the runway and take off in a continuous movement.

When unable to issue a take-off clearance:

CTL: Unable to Issue (designator) Departure (reasons);

CTL: Report When Ready [For Departure];

CTL: Are You Ready [For Departure]?;

CTL: Are You Ready For Immediate Departure?;

PIL: Ready;

CTL: Wait [reason].

Example 5:

☆ "United 684, taxi to holding point runway 17, report when ready for departure."

✛ "Taxi to holding point runway 17, United 684."

✛ "United 684, ready for departure"

☆ "United 684, wait, hold short of runway, aircraft short final."

When issuing a clearance to enter runway and wait take-off clearance:

CTL: Line Up [and Wait];

CTL: Line up Runway (number);

CTL: Line up. Be Ready For Immediate Departure;

CTL: (condition) Line up;

PIL: (condition) Lining up.

Example 6:

╋ "Japan Air 513, ready for takeoff."

☆ "Japan Air 513, line up and wait."

╋ "Line up and wait, Japan Air 513."

☆ "Japan Air 513, report airborne on departure frequency 126.4, cleared for takeoff."

When confirmation or readback of conditional clearance:

CTL: [That is] Correct [or I Say Again … (as appropriate)].

Example 7:

╋ "Pundong Ground, Air China 1259, confirm taxi to holding point runway 17 via taxiway D."

☆ "Air China 1259, Correct."

When issuing take-off clearance for more than one runway in use:

CTL: Cleared For Take-Off [Report Airborne];

CTL: Runway (number) Cleared for Take-off.

Example 8:

☆ "United 614, runway 17 cleared for take-off."

╋ "Cleared for take-off runway 17, United 614."

When the issued take-off clearance has not been complied with:

CTL: Take off Immediately or Vacate Runway [(instructions)];

CTL: Take off immediately or Hold Short of Runway.

Example 9:

☆ "United 321, are you ready for immediate departure?"

╋ "United 321, ready for departure."

☆ "United 321, take-off immediately or vacate runway."

╋ "United 321 rolling."

When cancel a take-off clearance:

CTL: Hold Position, Cancel Take-off. I Say Again Cancel Take-off (reasons);

PIL: Holding.

Example 10:

☆ "Air China 1425, hold position, I say again, cancel take-off. Aircraft on final. "

✛ "Air china 1425 holding. "

When stop a take-off after an aircraft has commenced take-off roll:

CTL: Stop immediately [(repeat aircraft call sign) Stop immediately];

PIL: Stopping.

Example 11:

✗ "Pudong Tower, China Southern 6234 rolling. "

☆ "China Southern 6234, stop immediately, I say again, China Southern 6234, stop immediately. "

✗ "China Southern 6234 stopping. "

When issuing clearance for helicopter operations:

CTL: Cleared for Take-off [From (location)] (present position, taxiway, final approach and take-off area, runway and number);

PIL: Request Departure Instructions;

CTL: After Departure Turn Right(or Left, or Climb)(instructions as appropriate).

Example 12:

🚁 "Pudong tower, G-CD, request departure instructions. "

☆ "G-CD, Cleared for take-off, after departure, turn right climb and maintain 300 meters, contact departure on 126. 4. "

🚁 "After departure, turn right climb and maintain 300 meters, contact departure on 126. 4 , G-CD. "

When issuing turn or climb instructions after take-off:

PIL: Request Right (or Left) Turn;

CTL: Right (or Left) Turn Approved;

CTL: Will Advise Later for Right (or Left) Turn;

CTL: Report Airborne;

CTL: Airborne (time);

CTL: After Passing (level) (instructions);

CTL: Continue Runway Heading (instructions);

CTL: Track Extended Centre line (instructions);

CTL: Climb Straight Ahead (instructions).

Example 13:

✗ "Pudong Tower, Singapore 3233 airborne."

☆ "Singapore 3233, continue runway heading, will advise later for left turn."

SUMMARY:

Aerodrome controller never issues takeoff clearance to pilot to permit take-off commencement until the preceding departing aircraft has:

- Crossed the end of the runway-in-use, or
- Has started a turn, or
- Until all preceding landing aircraft are clear of the runway-in-use.

7.4   Control of traffic in traffic circuit

Aerodrome controller issues the clearance to enter the traffic circuit to pilot whenever it is desired that the aircraft approach the landing area in accordance with current traffic circuits but traffic conditions cannot allow issuing landing clearance. Depending on the circumstances and traffic conditions, controller can clear the aircraft to join in the traffic circuit at any position.

Arriving aircraft executing instrument approach normally is cleared to land straight in unless visual maneuvering to the landing runway is required.

When aerodrome controller obverses an aircraft entering aerodrome traffic circuit without proper authorization, he/she generally permits the aircraft to land if its actions indicate that it so desires. If circumstances warrants, controller may instruct aircraft which are in contact with him/her to give way so as to remove as soon as possible the hazard which is introduced by such unauthorized operation. Controller cannot withhold permission to land indefinitely at any case.

In cases of emergency it may be necessary, in the interests of safety, for an aircraft to enter a traffic circuit and affect other landing without authorization from the controller. Controller has to recognize the possibilities of emergency action and render all assistance possible.

Priority of clearance to enter traffic circuit is as followings:

(a) an aircraft which anticipates being compelled to land because of factors affecting the safe operation of the aircraft (engine failure, shortage of fuel, etc.);

(b) hospital aircraft or aircraft carrying any sick or seriously injured persons requiring urgent medical attention;

(c) aircraft engaged in search and rescue operations; and

(d) other aircraft as may be determined by the state civil aviation authority.

When issuing instructions to join aerodrome traffic circuit:

PIL: [aircraft type] (position) (level) for Landing;

CTL: Join (position in circuit) (direction of circuit) (runway number) [Surface] Wind (direction and speed) (units) [Temperature [Minus] (number)] QNH (or QFE) (number) [(units)] [Traffic (detail)];

CTL: Make Straight-in Approach, Runway (number) [Surface] Wind (direction and speed) (units) [Temperature [Minus] (number)] QNH (or QFE) (number) [(units)] [Traffic (detail)];

PIL: (aircraft type) (position) (level) Information (ATIS identification) for Landing;

CTL: Join (position in circuit) [Runway (number)] QNH (or QFE) (number) [(units)] [Traffic (detail)].

Example 1:

✘ "Pudong Tower, United 56, 12 kilometers south 900 meters for landing."

☆ "United 56, make straight-in approach Runway 35, wind 340 degrees 7 meters per second, QNH 1031."

✘ "Straight-in approach Runway 35, QNH1031, United 56."

When issuing instructions to join right hand traffic circuit:

CTL: Join Right Hand (position in circuit) (runway number) [Surface] Wind (direction and speed) (units) [Temperature [Minus] (number)] QNH (or QFE) (number) [(units)] [Traffic (detail)];

Example 2:

✘ "Pudong Tower, China Eastern 126 Heavy, 10 kilometers north 700 meters, information Bravo for landing."

☆ "China Eastern 126, descend to circuit height, join right hand downwind Runway 35, QNH 1031."

✘ "Right hand downwind runway 35, QNH 1031, China Eastern 126."

When issuing instructions to aircraft in traffic circuit:

PIL: (position in circuit, e. g. Downwind/Final);

CTL：Number ... Follow (aircraft type and position) [additional instructions if required] .

Example 3：

✗ "Pudong Tower, China Southern 6019, downwind. "

☆ "China Southern 6019, number two, follow the Boeing 737 on final. "

✗ "Number two follow Boeing 737 on final, China Southern 6019. "

When issuing instructions for aircraft approach：

CTL：Make Short Approach;

CTL：Make Long Approach (or Extend Downwind);

CTL：Report Base (or Final, or Long Final);

CTL：Continue Approach [Prepare for Possible Go Around] .

Example 4：

✗ "Pudong ground, Air China 1052, Position downwind. "

☆ "Air China 1052, Make short approach, report base. "

✗ "Short approach, report base, Air China 1052. "

7. 5　Control of landing traffic

7. 5. 1　Landing clearance

Aerodrome controller may clear aircraft to land when there is reasonable assurance that the separation minima exists when the aircraft crosses the runway threshold, provided that a preceding landing aircraft has crossed the runway threshold. To reduce the potential for misunderstanding, the landing clearance includes the designator of the landing runway.

When issuing landing clearance：

CTL：Cleared to Land.

When issuing landing clearance for multiple runway operation：

CTL：Runway (number) Cleared to Land.

Example 1：

✗ "Pudong Tower, China Southern 684, long final. "

☆ "China Southern 684, Runway 35 cleared to land; wind 270 degrees 10 meters per second. "

When issuing landing clearance for special operations:

CTL: Cleared Touch and Go;

CTL: Make Full Stop.

Example 2:

✗ "Pudong Tower, Air China 592, request touch and go. "

☆ "Air China 592, cleared touch and go. "

When issuing instruction to an aircraft which request to fly past the control tower or other observation point for the purpose of visual inspection by persons on the ground:

PIL: Request Low Pass (reasons);

CTL: Cleared Low Pass [Runway (number)] [(altitude restriction if required) (go around instructions)];

Example 3:

✗ "Pudong Tower, United 624, Request low pass runway 17, unsafe left gear indication. "

☆ "United 624, cleared low pass Runway 17, descend no lower than 50 meters. After low pass, fly runway heading, climb 900 meters and maintain. "

When issuing landing clearance for helicopter operations:

PIL: Request Straight-in (or Circling Approach, Left (or Right) Turn to (location));

CTL: Make Straight-in (or Circling Approach, Left (or Right) Turn to (location, runway, taxiway, final approach and take-off area)) [Arrival (or Arrival Route) (number, name, or code)]. [Hold Short of (active runway, extended runway centre line, other)]. [Remain (direction or distance) from (runway, runway centre line, other helicopter or aircraft)]. [Caution (power lines, unlighted obstructions, wake turbulence, etc.)]. Cleared to land.

Example 4:

☞ "Pudong Tower, G-CD, request Straight-in to helicopter stand 5. "

☆ "G-CD, make straight-in to helicopter stand 5, remain distance from runway 17, caution jet blast, cleared to land. "

☞ "Straight-in to helicopter stand 5, remain distance from runway 17, G-CD. "

When issuing instruction to orbit an aircraft:

CTL: Circle the Aerodrome;

CTL: Orbit (Right, or Left) [From Present Position];

CTL: Make Another Circuit.

Example 5:

☆ "Hainan 789, extend downwind, number two, follow the B737 on four kilometer final."

✗ "B737 in sight, Hainan789."

☆ "Hainan 789, make one orbit right due to traffic on the runway, report again on final."

✗ "Orbiting right, Hainan 789."

✗ "Tower, Hainan 789, on final."

When issuing instruction for visual approach:

PIL: Request Visual Approach;

CTL: Cleared Visual Approach [Runway (number)].

Example 6:

✗ "Tower, CSN6901, runway in sight, request visual approach."

☆ "CSN6901, cleared visual approach runway 18L."

✗ "Cleared visual approach runway 18L, CSN6901."

Example 7:

☆ "CSN6901, turn right heading 150, report runway/preceding traffic in sight, and expect visual approach runway 18L."

✗ "Expect visual approach runway 18L, report runway/preceding traffic in sight, CSN6901."

✗ "Tower, CSN6901, runway/preceding traffic in sight."

☆ "CSN6901, follow the B757, cleared visual approach runway 18L. Caution wake turbulence."

When issuing instruction for visual approach vectoring:

CTL: Vectoring for Visual Approach Runnway (number), Report Field (or Runway) In Sight.

Example 8:

☆ "CCA1205, expect radar vectoring for visual approach runway 36R, report runway in sight."

✗ "expect radar vectoring for visual approach runway 36R, report runway in sight,

CCA1205. "

When issuing instruction for visual separation application:

CTL: Traffic, (clock position and distance), (direction) -bound, (type of aircraft), (intentions and other relevant information if applicable). Do you have it in sight?

CTL: Maintain Visual Separation.

Example 9 :

☆ "CCA1603, traffic 15km ahead of you, B777, report traffic in sight. "

✗ "Traffic in sight, CCA1603. "

☆ "CCA1603, maintain visual separation, caution wake turbulence. "

When issuing instruction for missed approach:

CTL: Go Around;

PIL: Going Around.

Example 10:

☆ "China Southern 6013, go around, aircraft on runway. "

✗ "China Southern 6013, going around. "

7.5.2 Landing and roll-out maneuvers

When necessary or desirable to expedite traffic, aerodrome controller may request a landing aircraft to:

(a) hold short of an intersecting runway after landing;

(b) land beyond the touchdown zone of the runway;

(c) vacate the runway at a specified exit taxiway;

(d) expedite vacating the runway.

When landing aircraft performs a specific landing and/or roll-out manoeuvre, the type of aircraft, runway length, location of exit taxiways, braking action on runway and taxiway, and prevailing weather conditions has to be considered. HEAVY aircraft cannot land beyond the touchdown zone of a runway.

If pilot considers that he/she is unable to comply with the requested operation, he/she must advise controller without delay.

When aerodrome controller considers it's necessary or desirable, e. g. low visibility conditions, he/she may instruct a landing or a taxiing aircraft to report runway vacated. Pilot has to report to controller when the aircraft is well clear of the runway.

When issuing instructions relating to communications after landing:

CTL：Contact Ground（frequency）；

CTL：When Vacated Contact Ground（frequency）.

Example 11：

☆ "Air China 1029, when vacated runway contact Ground 127.5."

✘ "Ground 127.5, Air China 1029."

When issuing instructions after landing:

CTL：Expedite Vacating；

CTL：Your Stand（or Gate）（designation）；

CTL：Take（or Turn）First（or Second, or Convenient）Left（or Right）and Contact Ground（frequency）.

Example 12：

☆ "China Southern 6115, Vacate runway, aircraft on short final, when vacated contact Ground 127.5."

✘ "Ground, China Southern 6115, runway vacated, request taxi instructions."

☆ "China Southern 6115, Ground, proceed south on taxiway D until reaching F intersection, turn first left, taxi to parking."

When issuing air taxi instructions after landing:

CTL：Air-taxi to Helicopter Stand（or）Helicopter Parking Position（area）；

CTL：Air-taxi to（or via）（location or routing as appropriate）[Caution（dust, blowing snow, loose debris, taxiing light aircraft, personnel, etc.）]；

CTL：Air Taxi via（direct, as requested, or specified route）to（location, heliport, operating or movement area, active or inactive runway）. Avoid（aircraft or vehicles or personnel）.

Example 13：

🚁 "Pudong Ground, G-CD, request air taxi to helicopter parking area."

☆ "G-CD, air taxi to helicopter parking area via taxiway T5, K4, caution construction in progress."

🚁 "Air taxi to helicopter parking area via taxiway T5, K4, G-CD."

SUMMARY:

Landing priority:

- aircraft that anticipates being compelled to land because of factors affecting the safe operation of the aircraft (engine failure, shortage of fuel etc.)
- Hospital aircraft or aircraft carrying any sick or seriously injured persons requiring urgent medical attention;
- Aircraft engaged in search and rescue operations; and
- Other aircraft which may be determined by state civil aviation authority.

7.6 Separation application and sequencing

Aerodrome controller provides aircraft in the traffic circuit separation minima outlined in Chapter 5 except that:

(a) aircraft in formation are exempted from the separation minima with respect to separation from other aircraft of the same flight;

(b) aircraft operating in different areas or different runways on aerodromes suitable for simultaneous landings or take-offs are exempted from the separation minima;

(c) no need to apply separation for aircraft operating under military necessity.

Aerodrome controller also has to provide sufficient separation between aircraft in flight in the traffic circuit to allow the spacing of arriving and departing aircraft.

Departures normally are cleared in the order in which they are ready for take-off, except that deviations may be made from this order of priority to facilitate the maximum number of departures with the least average delay. Factors which should be considered by the controller in relation to the departure sequence include, inter alia:

(a) types of aircraft and their relative performance;

(b) routes to be followed after take-off;

(c) any specified minimum departure interval between take-offs;

(d) necessity to apply wake turbulence separation minima;

(e) departure priority; and

(f) ATFM requirements.

7. 7　Example of a full flight operation radiotelephony communication

✛ "Pudong Delivery, United 65, destination Beijing, request ATC clearance. "

☆ "United 65, Pudong Delivery, cleared to BeiJing via flight plan route, request level change en-route, squawk 5151, contact ground 124. 5. "

✛ "Cleared to BeiJing via flight plan route, request level change en-route, squawk 5151, Ground 124. 5, United 65, good day. "

　…

✛ "Ground, United 65 heavy, gate 24, request push back. "

☆ "United 65 heavy, standby, expect two minutes delay due to Boeing 737 taxiing behind you. "

✛ "United 65 standby push back. "

☆ "United 65 pushback approved. "

✛ "Pudong Ground, United 65, gate 24, request start-up. "

☆ " United 65, Pudong Ground, expect departure at 49, start up at own discretion. "

✛ "Expect departure at 49, United 65. "

✛ "United 65, pushing back. "

✛ "Pudong Ground, United 65, request taxi. "

☆ "United 65, taxi via taxiway C to holding point runway 17, caution, construction work next to taxiway. "

✛ "Taxi via taxiway C to holding point runway 17, United 65. "

☆ "United 65, hold your position, give way to the Boeing 747 from left to right. "

✛ "Hold position, United 65. "

☆ "United 65, continue taxi to holding point runway 17, Contact tower on 118. 7. "

✛ "Tower 118. 7, good day, United 65 . "

　…

✛ "Pudong Tower , United 65 , ready for departure. "

☆ "United 65, Pudong Tower, report the airbus on final in sight. "

✛ "Airbus in sight, United 65. "

☆ "United 65, after the landing Airbus, line up and wait. "

✛ "United 65, wilco. "

☆ "United 65, after takeoff continue runway heading, climb 600 meters and maintain, caution wake turbulence, cleared for takeoff, wind 320 degrees 3 meters per second. "

✈ "United 65 rolling. "

✗ "Tower, United 65, airborne. "

☆ "United 65, radar identified. Maintain present heading, climb 600 meters, Contact departure 126. 7. "

✗ "Climb 600meters, departure 126. 7 United 65 Heavy, Good day. "
...

✗ "Beijing Tower, United 65, 10 kilometers south, 600 meters, information Delta for landing. "

☆ "United 65, Beijing Tower, descend to circuit height, join right hand downwind Runway 35, QNH 1031. "

✗ "Right hand downwind Runway 35, QNH 1031, United 65. "

☆ "United 65, follow the Boeing 737 on final, number two to land, report turning base. "

✗ "Follow the Boeing 737 on final, number two to land, United 65. "

✗ "United 65, turning base, B737 on final in sight. "

☆ "United 65, cleared to land, wind 340 degrees 3 meters per second. "

✗ "Cleared to land, United 65. "

☆ "United 65, Expedite vacate runway, aircraft on final, reported vacated. "

✈ "United 65, runway vacated. "

☆ "United 65, contact ground 125. 7, good day. "
...

✈ "Beijing Ground, United 65, runway vacated, request taxi instructions. "

☆ "United 65, Beijing Ground, proceed south on taxiway D until reaching F intersection, turn left, taxi to parking. "

7. 8   Common used methods for conflict resolution

This section includes the basic principle and common methods for resolving traffic conflict during aerodrome control service, and extends the previous theoretical lessons.

7. 8. 1   Air traffic controller conflict detection and resolution process

Air traffic control is a highly knowledge and information intensive activity. From the aspects of the controller, the main objective of air traffic control is to acquire and maintain safe separation between aircraft in accordance with relevant rules by issuing proper instructions to the pilots. Thus conflict is defined as the situation of loss of minimum safe separation between two aircraft. In other words, the distance between aircraft violates a criterion defining what is considered undesirable.

Conflict detection and resolution are core tasks of the controller. In process of conflict detection, the objective is to evaluate conflict probability over a certain future horizon starting from the current positions and flight plans of the aircraft. In conflict resolution the objective is to calculate suitable maneuvers to avoid a predicted conflict.

Humans are an essential element in this process due to their ability to integrate information and make judgments. Controller manages the complexity of air traffic scenarios by disassembling complex problems into simple ones, by monitoring some key parameters which are crucial to the development of safe traffic. This process could occur in different ways according to different controllers. Ground-based control essentially consists of progress of aircraft represented by points on a flat display screen. The simple nature of the data available means that the controllers themselves are required to build and maintain a "mental picture" of extrapolated 4D traffic based on experience and other knowledge. Controller mentally compares every pair of predicted trajectories to determine whether any pair of aircraft will pass within the minimum permitted separation - in which case he/she is required to intervene in some way to resolve the potential conflict.

In conflict resolution, there are often many potential resolutions, but there are two principal problems for a controller are the same—the first is one of trying to define the possible solutions, and the second is that of trying to derive an optimal resolution. The process of detection and resolution can be illustrated as Fig 7.3.

Strategy of a controller is the key of conflict resolution behavior. They are the very conscious and resulting from the controllers' thinking and experience. The above process leads the controller to select a resolution from certain strategies. The final resolution will be up to the controller based on his/her cognitive experience and instructions given to the pilot by the controller will be in the following dimensions:

(a) Lateral (turn one or both aircraft)

(b) Vertical (climb or descend aircraft)

(c) Speed (increase or reduce)

The factors which will affect controller's strategies and his/her final solution include the followings:

(a) number of problems present at any given time

(b) number of uncertainties

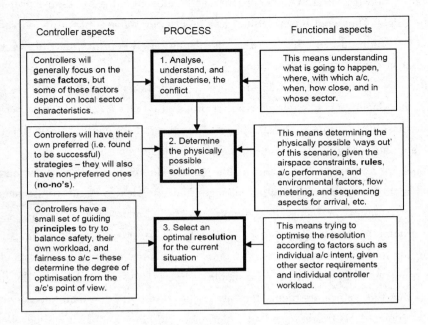

Fig 7. 3   Controller conflict detection and resolution process

(c) willingness to consider a low probability as irrelevant (risk tolerance)

(d) management of their own cognitive resources

7. 8. 2   Common used methods for aerodrome traffic conflict resolution

Aerodrome control is the start or end of air traffic control service and perhaps is one of the most difficult scenarios in ATC carrier. The management of traffic, for the aerodrome controller, includes tasks such as responsible for the pre-departure coordination of flight plan information with pilots, control and coordinates the movement of aircraft on the airport surface, determining and sequencing departure aircraft, responsible for controlling aircraft landings and departures, etc. The characteristic of aerodrome traffic conflict resolution is that the controller need to consider both airborne and on ground traffic together with flight plan and the airspace is very limited. These will confine the methods which aerodrome controller could use in traffic conflict resolution. Hereunder the methods which are common used by aerodrome controller has been listed.

7. 8. 2. 1   For arriving aircraft

Generally, approach controller shall have sequenced the arriving aircraft before handover to tower controller. The approach controllers shall handover the aircraft in accordance with the LOA, or request an acceptance approval from the tower. Tower need not rearrange the sequence except that an emergence situation occurs.

7. 8. 2. 2   For departing aircraft

Besides issuance of the release clearance, the tower may resolve the conflict for runway occupancy by holding the departing aircraft at the holding point or even on the runway. Hold the aircraft at the holding point generally is used for the assurance of the separation between a preceding landing aircraft and the departing one; Hold the aircraft on the runway can be used for wake turbulence behind a succeeding departure.

7. 8. 2. 3   For traffic in traffic circuit

(a) choosing an appropriate joining point and procedure to join the traffic circuit

To resolve the confliction in traffic circuit, tower controller may designate an appropriate joining point and entering procedure for arriving aircraft from different direction to ensure the separation between aircraft in traffic circuit. The designated joining point with entering procedure shall be determined based on aircraft entrance direction and consider aircraft performance and convenience for the pilot. The common used joining point with its entrance procedure is illustrated as the Fig 6-1.

(b) maneuvering

Maneuvering is very effective way to delay the aircraft provided the airspace is allowed. Maneuvering shall consider aircraft performance and with consent of the pilot. It can be made in downwind, base and final. Besides traffic circuit, aircraft with straight-in approach may also be instructed to make a maneuvering for separation.

(c) extend downwind

This method generally is used for delaying the aircraft on downwind to resolve the confliction between the traffic on downwind and final. The elapsing time usually is an increment with 30 seconds and not exceeds maximum 2 minutes.

(d) make a short final

This method generally is used for shortening the flight time for landing aircraft in traffic circuit to ensure the separation from the behind approaching or departing aircraft. The flight time which can be shortened is depend on aircraft performance and may be very limited.

(e) make another circuit

When methods such as maneuvering or extending downwind can not solve confliction, the controller can delay the landing aircraft by instructing the aircraft make another circuit. In fact, it means to hold the aircraft over the aerodrome for a long period and the vertical separation may also be applied for by this method.

(f) apply visual separation

By applying visual separation, controller may reduce horizontal separation between two aircraft in traffic circuit, but this does not relieve the controller's responsibility to ensure wake turbulence or converging confliction.

(g) speed adjustment

Sometimes speed adjustment may be used for ensuring separation in traffic circuit even though it may not be very an effective way. Tower controller shall enquire the pilot if the aircraft performance is allowed to do so.

Exercise

I. Explain the following definitions.

(1) back track

(2) at own discretion

(3) line up

(4) hold short

(5) ATC clearance

(6) expedite cross

(7) holding position runway 07

(8) air taxi

(9) short approach

(10) touch and go

(11) low pass

(12) make an orbit

(13) visual approach

(14) conflict

(15) maneuver

(16) 4D trajectories

(17) short final

(18) cognitive experience

(19) converging confliction

II. Answer the following questions.

(1) Under what circumstance controller cannot clear aircraft to line up and hold on the approach end of a runway-in-use?

(2) Give an example of ATC clearance.

(3) What contents shall be contained in the standard clearance for a departing aircraft?

(4) Which items in ATC clearances and instructions shall be read back to aerodrome controller by the pilot?

(5) What rules shall the vehicles comply with on the maneuvering area?

(6) Write down an example of issuing instructions to join the aerodrome traffic circuit using standard phraseology.

(7) Describe briefly the process of controller conflict detection and resolution.

(8) What's the key element for controller conflict resolution behavior? What factors may affect it?

(9) What are common used methods for aerodrome traffic conflict resolution?

Ⅲ. Find out the inappropriate point in the following controller-pilot communication and give a correction.

(1) PIL: TianJin tower, C-G1, good morning, 8 kilometers east of runway 600 meters information C for landing.
CTL: C-G1, straight in approach runway 27, report on final.

(2) PIL: TianJin tower, C-G2, ready for departure.
CTL: C-G2, clear for takeoff 2 minutes later due to wake turbulence.

(3) CTL: C-G3, clear to land, surface wind 330 degrees 02 meters per seconds, QNH1016.
PIL: C-G3 roger.
CTL: C-G3, report runway vacated.

Ⅳ. **Write down the appropriate taxi instruction issued to aircraft N-CD based on the expected taxi route with dash line in the following figure. If N-CD expects to depart runway** 09, **circle out the point where departure information shall be issued and takeoff clearance shall be issued.**

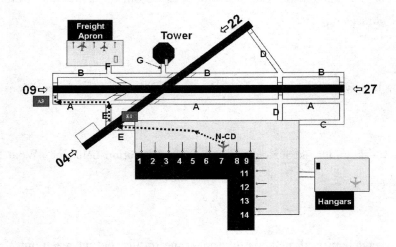

Ⅴ. **Practice the pilot-controller communication in pairs when aircraft N-CD request join the traffic circuit RWY27 while it approaches aerodrome from different position in the following figure.**

(1) from the train station (2) from the school (3) from the stadium
(4) from the hospital (5) from the church

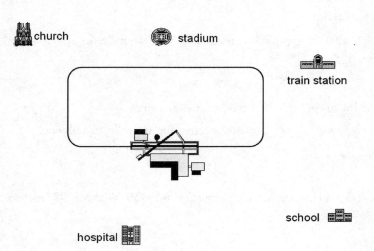

VI. You have the traffic scenario as the following figure. N-CF1 is number 1 to land and aerodrome has issued landing clearance. N-CF2 is on downwind and followed by N-CF3. G-HE is at holding position of runway 27 and ready for departure. Answer the following questions.

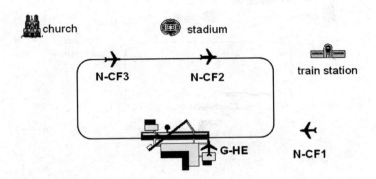

(1) If N-CF3 is faster than N-CF2 and is catching up with it, as an aerodrome controller, what methods can be used to ensure the landing sequences? Why? What instruction shall you pass to pilots? Practice the pilot-controller communication in pairs and then swap the roles.

(2) Suppose N-CF1 is 5 NM from the threshold while G-HE is requesting for departure, do you think there is enough time to get G-HR airborne? What instructions shall you pass to pilots of N-CF1 and G-HE? Practice in pairs for this situation.

(3) Suppose N-CF2 encounters an engine failure problem and need priority number 1 to land, what methods can be used to resolve this situation? What circumstance may N-CF2 result in? Practice in pairs for this situation and then swap the roles.

VII. Answer the following questions based on traffic scenario in following figure（runway in use 27）.

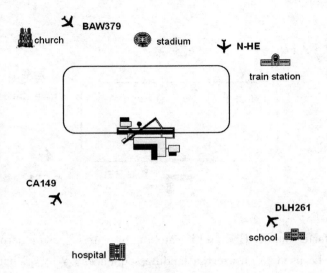

（1）According to your strategy, what's your intended landing sequence? and why?

（2）What methods can you use to ensure the sequence? Discuss with your partner and practice in pairs.

（3）Suppose CA149 encounters engine fire warning emergency and request priority landing, what's your landing sequence then and how do you resolve this situation and practice in pairs.

Chapter 8

Local instructions

An aerodrome local instruction is part of content included in AIP, Part 3-Aerodromes (AD) by the contracted state. It constitutes the basic information of an aerodrome for permanent information and long duration temporary changes.

8. 1   An example of local instruction (TianJin/Binhai airport)

The local instruction of TianJin/Binhai airport hereunder is given in this textbook has been attached in the following pages.

AD 2.1 Aerodrome location indicator and name

ZBTJ—TIANJIN/Binhai

ZBTJ AD 2.2 Aerodrome geographical and administrative data

1	ARP coordinates and site at AD	N39°07.5' E117°20.8' Center of RWY
2	Direction and distance from city	095° GEO, 13.3km from city center square
3	Elevation/Reference temperature	4m/ 32.0°C(JUL)
4	MAG VAR/Annual change	6°W(1975)/-
5	**AFS** **AD** administration, address, telephone, telefax, **AFS**	Tianjin Municipal Administration of CAAC Tianjin Binhai International Airport, Dongli District, Tianjin 300300, China TEL: 86-22-24901114 FAX: 86-22-24394233 AFS: ZBTJYFYX
6	Types of traffic permitted	IFR/VFR
7	Remarks	Nil

ZBTJ AD 2.3 Operational hours

1	AD Administration (AD operational hours)	HS or O/R
2	Customs and immigration	HS or O/R
3	Health and sanitation	HS or O/R
4	AIS Briefing Office	HS or O/R
5	ATS Reporting Office (ARO)	HS or O/R
6	MET Briefing Office	HS or O/R
7	ATS	HS or O/R
8	Fuelling	HS or O/R
9	Handling	HS or O/R
10	Security	HS or O/R
11	De-icing	HS or O/R
12	Remarks	Nil

ZBTJ AD 2. 4 Handling services and facilities

1	Cargo-handling facilities	2 Platform lift(13 tonnes, 7 tonnes)
2	Fuel/oil types	Nr.3 jet fuel --
3	Fuelling facilities/capacity	Refueling truck and hydrant cart: 1000-1200 litres/ min
4	De-icing facilities	4 snow ploughs, 1 de-icer
5	Hangar space for visiting aircraft	Nil
6	Repair facilities for visiting aircraft	Line maintenance available for aircraft type below B767 or A300(include) on request
7	Remarks	Nil

ZBTJ AD 2.5 Passenger facilities

(Omitted)

ZBTJ AD 2.6 Rescue and fire fighting services

(Omitted)

ZBTJ AD 2.7 SEASNAL AVAILABILITY-CLEARING

(Omitted)

ZBTJ AD 2.8 Aprons, taxiways and check locations data

1	Apron surface and strength	Surface:	Cement concrete
		Strength:	PCN 60/R/B/X/T(stands 1-17, 51-62)
			PCN 81/R/B/X/T(stands 30-39)
2	Taxiway width, surface and strength	Width:	23m: A. E. J. Z; 18m: B. D; 25m: C; 24m: F·; 34m: G. H.
		Surface:	Asphalt: A. B. C. D. E. Z
			Cement concrete: F. G. H. J
		Strength:	PCN 83/F/B/W/T (A. B. C. D. E)
			PCN 60/R/B/X/T (F. J)
			PCN 81/R/B/X/T (G. H)
			PCN 83/R/B/W/T (Z)
3	Taxiway width, surface and strength	Nil	
4	VOR/INS checkpoints	Nil	
5	Remarks	Nil	

ZBTJ AD 2.9 Surface movement guidance and control system and markings

1	Use of aircraft stand ID signs, TWY guidelines and visual docking/parking guidance system of aircraft stands	Taxiing guidance signs at all intersections of RWY and TWY and at all holding positions. Guide lines at apron. Follow-me car.	
2		RWY markings	THR, RWY designation, TDZ, center line, edge line, aiming point
		RWY lights	Center line, edge line, THR, RWY end, wing bar
		TWY markings	Center line, taxi holding position
		TWY lights	Edge line
3	Stop bars	Nil	
4	Remarks	Nil	

ZBTJ AD 2.10   Aerodrome obstacles

Refer to AD IAC, No significant obstacles in the take-off flight path area.

ZBTJ AD 2.11 Meteorological information provided

1	Associated MET Office	Tianjin Aerodrome MET Office
2	Hours of service, MET Office outside hours	H24 --
3	Office responsible for TAF preparation, Periods of validity	Tianjin Aerodrome MET Office 10 HR, 24 HR
4	Type of landing forecast, Interval of issuance	Trend 30 MIN
5	Briefing/consultation provided	P,T
6	Flight documentation, Languages used	Chart, International MET Codes, Abbreviated Plain Language Text Ch, En
7	Charts and other information available for briefing or consultation	Synoptic charts, significant weather charts, upper W/T charts, satellite and radar material, AWOS real-time data
8	Supplementary equipment available for providing information	MET Service Terminal, FAX
9	ATS units provided with information	Tianjin TWR, Tianjin APP
10	Additional information	Nil

ZBTJ AD 2.12 Runway physical characteristics

Designations RWY NR	TURE& MAG BRG	Dimensions of RWY (m)	Strength(PCN) and surface of RWY and SWY	THR Coordinates	THR elevation and highest elevation of TDZ of precision APP RWY
1	2	3	4	5	6
16	154° GEO 160° MAG	3200×50	83/F/B/W/T Asphalt	Nil	THR 2.4m --
34	334°GEO 340°MAG	3200×50	83/F/B/W/T Asphalt	Nil	THR 2.4m --

Slope of RWY-SWY	SWY dimensions (m)	CWY dimensions (m)	Strip dimensions (m)	OFZ	Remarks
7	8	9	10	11	12
RWY 16/34 +0.02%(280) 0%(920) +0.01%(400) +0.11%(703) -0.08%(897)	60×50 60×50	200×190 200×190	3320×300 3320×300	Nil Nil	Nil Nil

ZBTJ AD 2.13 Declared distances

RWY Designator	TORA (m)	TODA (m)	ASDA (m)	LDA (m)	Remarks
1	2	3	4	5	6
16	3200	3400	3260	3200	Nil
34	3200	3400	3260	3200	Nil

ZBTJ A.D 2.14   Approach and runway lighting

RWY Designator	APCH LGT type LEN INTST	THR LGT Color WBAR	VASIS (MEHT) PAPI	TDZ LGT LEN	RWY Center line LGT LEN, spacing, color, INTST	RWY Edge LGT LEN, spacing, color,INTST	RWY End LGT color, WBAR	SWY LGT LEN, color
1	2	3	4	5	6	7	8	9
16	CAT I 900m LIH	Green Yes	PAPI Left/3°	Nil	3200m ** spacing 30m	3200m*** Spacing 50m	Red Green	Nil
34	CAT I* 900m LIH	Green Yes	PAPI Left/3°	Nil	3200m ** spacing 30m	3200m*** Spacing 50m	Red Green	Nil

Remarks: * SFL(beyond 300m FM THR)

** 0-2200m White LIH, 2200-2900m Red/White alternative LIH, 2900-3200m Red LIH

*** 0-2700m White LIH, 2700-3200m Yellow LIH

ZBTJ AD 2.15 Other lighting, secondary power supply

1	ABN/IBN location, characteristics and hours of operation	Nil
2	LDI location and LGT, Anemometer location and LGT	White landing 'T', lighted: see AD Chart
3	TWY edge and center line lighting	Edge line lights available
4	Secondary power supply/switch-over time	Standby power supply available/ 5sec
5	Remarks	Nil

ZBTJ AD 2.16 Helicopter landing area

Nil

ZBTJ AD 2.17 ATS airspace

1	Designation and lateral limits	Tianjin tower control area N390900E1171400—N391600E1171400— N391800E1172200—N385800E1174000— N385800E1172000—N390900E1171400
2	Vertical limits	SFC-600m MSL
3	Airspace Classification	To be developed
4	ATS unit call sign, Languages	Tianjin Tower Ch, En
5	Transition altitude	3000m
6	Remarks	Nil

ZBTJ AD 2.18 ATS communication facilities

Service Designation	Call sign	Frequency (MHZ)	Hours of operation	Remarks
1	2	3	4	5
APP	Tianjin Approach	127.9 (120.9)	H24	Nil
TWR	Tianjin Tower	118.2 (130.0)	H24	Nil
GND	Ramp Control	121.75	HO	Nil
ATIS		126.4	H24	Nil

ZBTJ AD 2.19 Radio navigation and landing aids

Type of aid	ID	Frequency	Hours of operation	Antenna Site coordinates	DME antenna elevation	Remarks
1	2	3	4	5	6	7
VOR/DME	TAJ	112.1 MHZ CH58X	HO	N39° 06.8' E117° 21.3'	6m	
LOM 16	JS	452 KHZ	HO	340° MAG/ 4100m FM THR RWY 16	4.1m	
LMM 16	J	517 KHZ	HO	340° MAG/ 1100m FM THR RWY 16		
ILS 16 LLZ	IJS	110.9 MHZ	HO	160° MAG/ 400m FM end RWY 16		Beyond 26° rightside of front course U/S
GP 16		330.8 MHZ	HO	132m W of RCL 335m FM THR 16		Angle 3°, RDH 15.6m Coverage 25km
DME 16	IJS	CH46X (110.9 MHZ)	HO		11m	Co-located with GP
LO 34	CG	339 KHZ	HO	N39° 04.4' E117° 22.7'		160° MAG/ 4000m FM THR RWY 34; Bearing 286° U/S
LMM 34	C	419 MHZ	HO	160° MAG/ 1100m FM THR RWY 34		
OM 34		75 MHZ	HO	160° MAG/ 7795m FM THR RWY 34		
ILS 34 LLZ	ICG	110.5 MHZ	HO	340° MAG/ 350m FM end RWY 34		
GP 34		329.6 MHZ	HO	132m W of RCL 335m FM THR		Angle 3°, RDH 16.6m Coverage 25km
DME 34	ICG	CH42X (110.5 MHZ)	HO		11m	Co-located with GP

AIP CHINA   **ZBTJ AD2-8**

ZBTJ AD 2.20 Local traffic regulations

1. AD operation regulations

Each and every training flight or technical test flight shall be filed in advance and conducted only after clearance has been obtained from ATC.

2. Use of runways and taxiways

a. Follow-me vehicle service is available via Tower Control;

b. 180° turnaround on TWY is forbidden for all aircraft.

3. Use of aprons and parking stands

a. Engine run-ups are subject to Ground Control clearance, and shall be carried out at a designated location;

b. Wingspan limits for aircraft parking on the stands:

64.9 m: Nr. 1, 7, 11, 37, 38, 39;

47.6 m: Nr. 2, 14, 15, 33, 34, 35, 36;

34.3 m: Other Stands.

4. CAT II/III operations at AD:

Nil

5. warning

Nil

7. Helicopter operation restrictions and helicopter parking/docking area

Nil

ZBTJ AD 2.21 Noise abatement procedures

Nil

ZBTJ AD 2.22 Flight procedures

1. General

Flights within Tianjin Approach Control Area and Tower Control Area shall operate under IFR unless special clearance has been obtained from Tianjin Approach Control or Tower Control.

2. Traffic circuits

Traffic circuits shall be made to the east of RWY, at the altitude of 300m for aircraft CAT A/B, and 300m-500m for aircraft CAT C/D.

3. IFR flight procedures within Tianjin Approach Control Area and Tower Control Area

a. Strict adherence is required to the relevant arrival/departure procedures published in the aeronautical charts and the relevant regulations published in subsection ENR2.2.1. Aircraft may, if necessary, hold or maneuver on an airway, over a navigation facility or a fix designated by ATC;

b. Flight altitudes within the area of N3909.0 E11714.0-N3916.0 E11714.0-N3918.0 E11722.0-N3913.0 E11727.5-N3909.0 E11714.0 are restricted at or below 800m unless prior applications have been made.

4. Radar procedures

a. Radar control within Tianjin APP has been implemented. The minimum horizontal radar separation is 10km and the minimum vertical radar separation is 300m for aircraft within Tianjin APP;

b. Radar vectoring and sequencing:

Normally, aircraft will be vectored and sequenced from Dawangzhuang VOR(VYK), Shigezhuang NDB(VM) or transfer of control points to the appropriate final approach track or to the time when RWY is in sight. Instructions about radar vectors, ascent/descent altitudes or speed adjustment will be issued for spacing and separating the aircraft so that stipulated radar intervals and wake intervals are maintained, taking into account aircraft characteristics or control regulations.

5. Radio communication failure procedures

Nil

6. Procedures for VFR flights

Nil

7. VFR route

Nil

8. Visual reference point

Nil

9. Other regulations

Nil

ZBTJ AD 2.23 Other Information

1. Fuel dumping area

Lateral limits: N38 58.0E117 58.0— N38 35.0E119 24.0—N38 24.0E119 19.0— N38 49.0E117 55.0—N38 58.0E117 58.0

Vertical limits: above 4000m

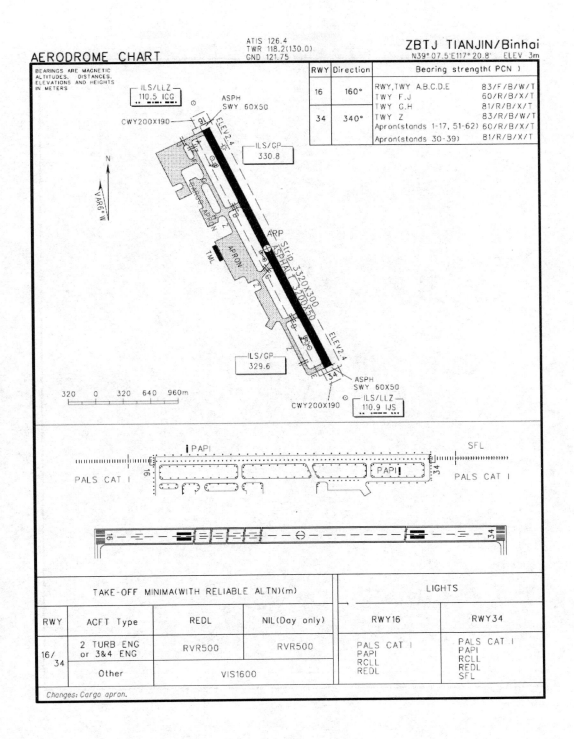

AERODROME CHART

ATIS 126.4
TWR 118.2(130.0)
GND 121.75

ZBTJ TIANJIN/Binhai
N39°07.5'E117°20.8'   ELEV 3m

BEARINGS ARE MAGNETIC.
ALTITUDES, DISTANCES,
ELEVATIONS AND HEIGHTS
IN METERS

RWY	Direction	Bearing strength(PCN)	
16	160°	RWY,TWY A.B.C.D.E	83/F/B/W/T
		TWY F.J	60/R/B/X/T
		TWY G.H	81/R/B/X/T
34	340°	TWY Z	83/R/B/W/T
		Apron(stands 1-17, 51-62)	60/R/B/X/T
		Apron(stands 30-39)	81/R/B/X/T

ILS/LLZ
110.5 ICG

ASPH
SWY 60x50

CWY200x190

ELEV2.4

ILS/GP
330.8

N
VAR6°W

Cargo APRON

APRON

TNL

ARP
Strip 3320X300
ASPH 3200X50

ELEV2.4

ILS/GP
329.6

34

ASPH
SWY 60x50

320 0 320 640 960m

CWY200x190

ILS/LLZ
110.9 IJS

PAPI   SFL

PALS CAT I   16   PAPI   34   PALS CAT I

16   34

TAKE-OFF MINIMA(WITH RELIABLE ALTN)(m)

LIGHTS

RWY	ACFT Type	REDL	NIL(Day only)	RWY16	RWY34
16/34	2 TURB ENG or 3&4 ENG	RVR500	RVR500	PALS CAT I PAPI RCLL REDL	PALS CAT I PAPI RCLL REDL SFL
	Other	VIS1600			

Changes: Cargo apron.

AIRCRAFT-PARKING
CHART-ICAO

ATIS 126.4
TWR 118.2(130.0)
GND 121.75

ZBTJ TIANJIN/Binhai

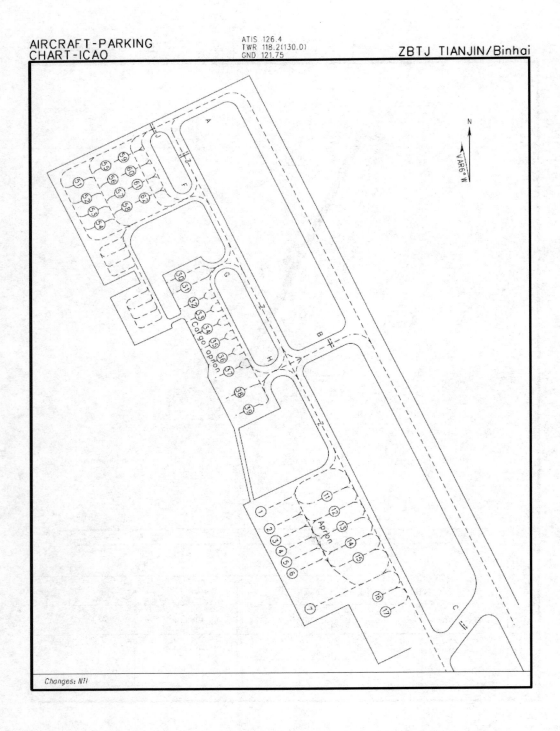

STANDARD DEPARTURE
CHART-INSTRUMENT

VAR6° W

ATIS 126.4
APP 127.9(120.9)
TWR 118.2(130.0)

ZBTJ TIANJIN/Binhai
RWY16

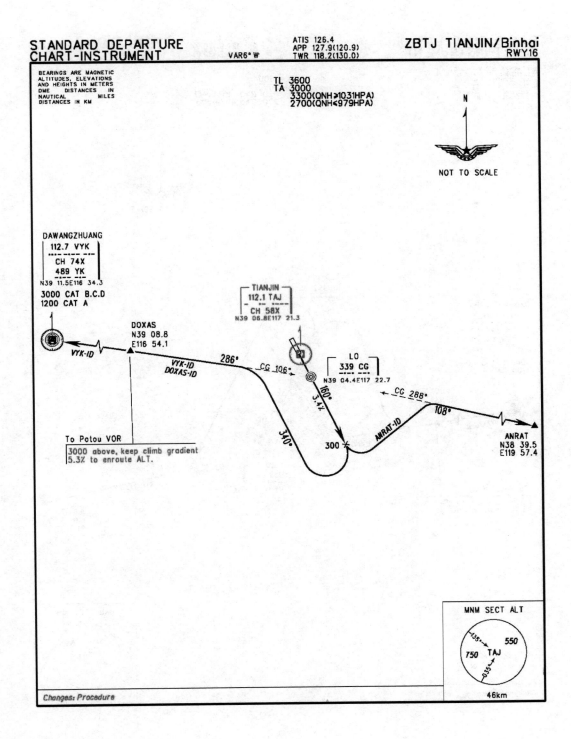

BEARINGS ARE MAGNETIC
ALTITUDES, ELEVATIONS
AND HEIGHTS IN METERS
DME DISTANCES IN
NAUTICAL MILES
DISTANCES IN KM

TL 3600
TA 3000
3300(QNH>1031HPA)
2700(QNH<979HPA)

N

NOT TO SCALE

DAWANGZHUANG
112.7 VYK
CH 74X
489 YK
N39 11.5E116 34.3
3000 CAT B.C.D
1200 CAT A

VYK-ID

DOXAS
N39 08.8
E116 54.1

VYK-ID
DOXAS-ID

286°

TIANJIN
112.1 TAJ
CH 58X
N39 06.8E117 21.3

CG 106°

LO
339 CG
N39 04.4E117 22.7

160°
3.4%

CG 288°

108°

340°

300°

ANRAT-ID

ANRAT
N38 39.5
E119 57.4

To Potou VOR
3000 above, keep climb gradient
5.3% to enroute ALT.

MNM SECT ALT
135° 550
750 TAJ
315°
46km

Changes: Procedure

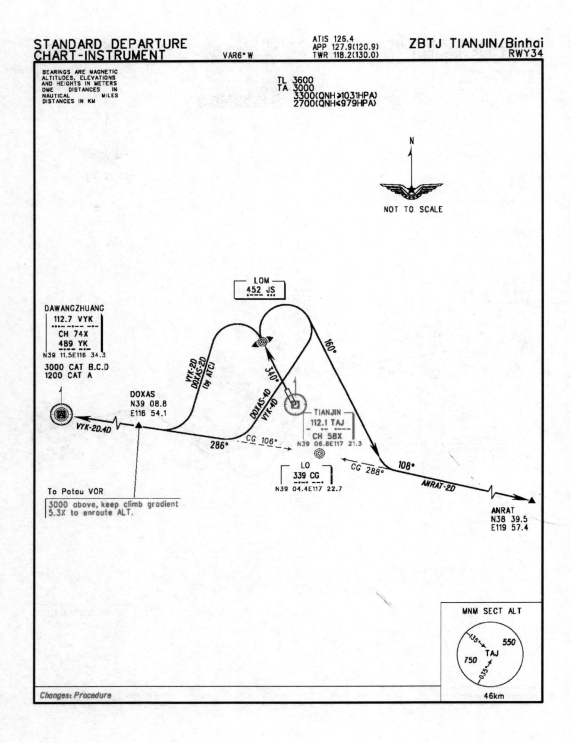

STANDARD DEPARTURE CHART-INSTRUMENT

VAR6° W

ATIS 126.4
APP 127.9(120.9)
TWR 118.2(130.0)

ZBTJ TIANJIN/Binhai
RWY34

BEARINGS ARE MAGNETIC
ALTITUDES, ELEVATIONS
AND HEIGHTS IN METERS
DME DISTANCES IN
NAUTICAL MILES
DISTANCES IN KM

TL 3600
TA 3000
    3300(QNH≥1031HPA)
    2700(QNH<979HPA)

N

NOT TO SCALE

LOM
452 JS

DAWANGZHUANG
112.7 VYK
CH 74X
489 YK
N39 11.5E116 34.3
3000 CAT B.C.D
1200 CAT A

VYK-2D
DOXAS-2D
(by ATC)

DOXAS
N39 08.8
E116 54.1

DOXAS-4D
VYK-4D

340°

160°

TIANJIN
112.1 TAJ
CH 5BX
N39 06.8E117 21.3

VYK-2D.4D

286°

CG 106°

LO
339 CG
N39 04.4E117 22.7

CG 288°

108°

ANRAT-2D

To Potou VOR

3000 above, keep climb gradient
5.3% to enroute ALT.

ANRAT
N38 39.5
E119 57.4

MNM SECT ALT

135°
550
TAJ
750
035°

46km

Changes: Procedure

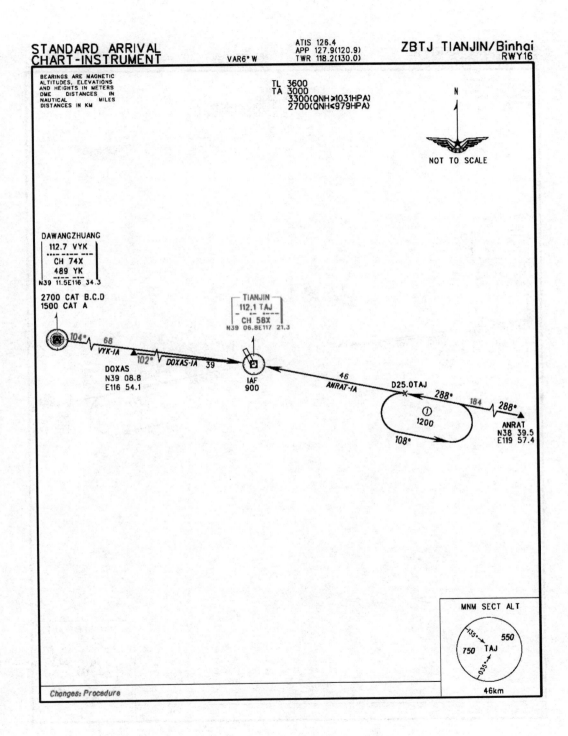

STANDARD ARRIVAL
CHART-INSTRUMENT

VAR6°W

ATIS 126.4
APP 127.9(120.9)
TWR 118.2(130.0)

ZBTJ TIANJIN/Binhai
RWY16

BEARINGS ARE MAGNETIC
ALTITUDES, ELEVATIONS
AND HEIGHTS IN METERS
DME   DISTANCES   IN
NAUTICAL   MILES
DISTANCES IN KM

TL 3600
TA 3000
   3300(QNH≥1031HPA)
   2700(QNH<979HPA)

N

NOT TO SCALE

DAWANGZHUANG
112.7 VYK
CH 74X
489 YK
N39 11.5E116 34.3

2700 CAT B.C.D
1500 CAT A

TIANJIN
112.1 TAJ
CH 58X
N39 06.8E117 21.3

104°   68
VYK-1A
102°   DOXAS-1A   39

DOXAS
N39 08.8
E116 54.1

IAF
900

46
ANRAT-1A

D25.0TAJ
×

288°   184

288°

①
1200

108°

ANRAT
N38 39.5
E119 57.4

MNM SECT ALT

135°
035°
550
750   TAJ

46km

Changes: Procedure

STANDARD ARRIVAL CHART-INSTRUMENT

VAR6°W

ATIS 126.4
APP 127.9(120.9)
TWR 118.2(130.0)

ZBTJ TIANJIN/Binhai
RWY34

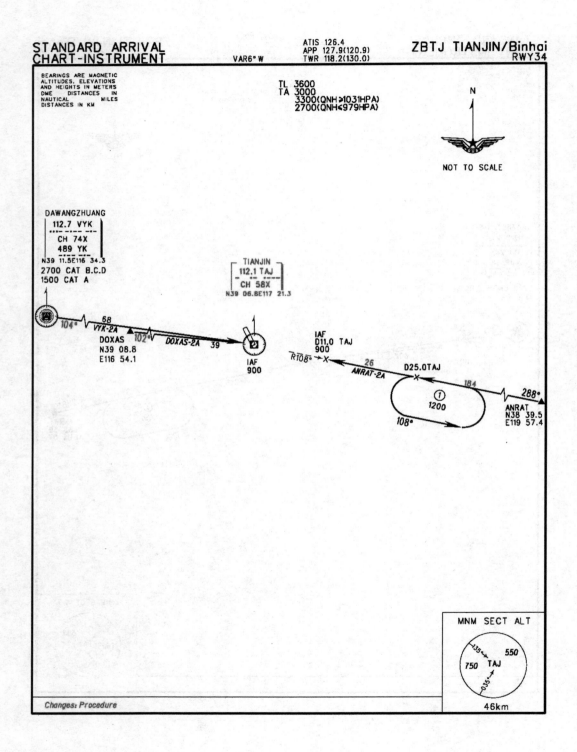

BEARINGS ARE MAGNETIC
ALTITUDES, ELEVATIONS
AND HEIGHTS IN METERS
DME DISTANCES IN
NAUTICAL MILES
DISTANCES IN KM

TL 3600
TA 3000
3300(QNH≥1031HPA)
2700(QNH<979HPA)

N

NOT TO SCALE

DAWANGZHUANG
112.7 VYK
CH 74X
489 YK
N39 11.5E116 34.3
2700 CAT B.C.D
1500 CAT A

TIANJIN
112.1 TAJ
CH 58X
N39 06.8E117 21.3

104°
68
VYK-2A
DOXAS 102° DOXAS-2A 39
N39 08.8
E116 54.1

IAF
900

IAF
D11.0 TAJ
900

R108° X
26
ANRAT-2A
D25.0TAJ
X
184

288°

① 1200
ANRAT
N38 39.5
E119 57.4

108°

MNM SECT ALT

135 550
750 TAJ
035

46km

Changes: Procedure

· 156 ·

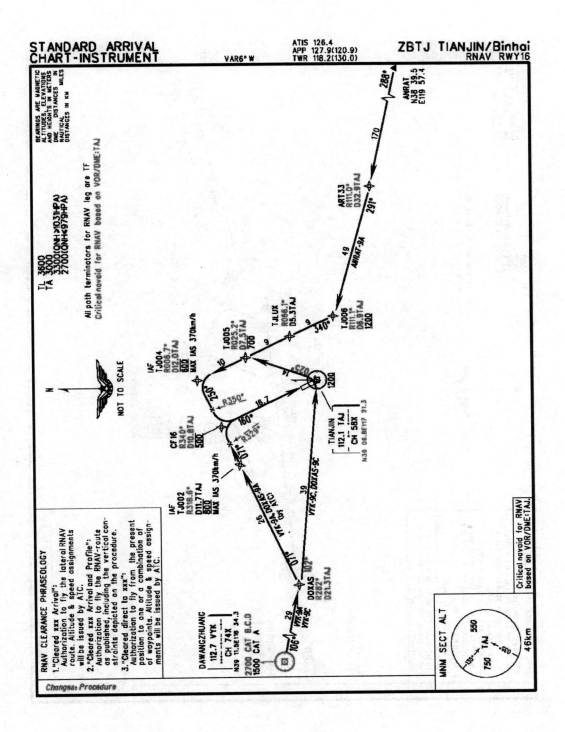

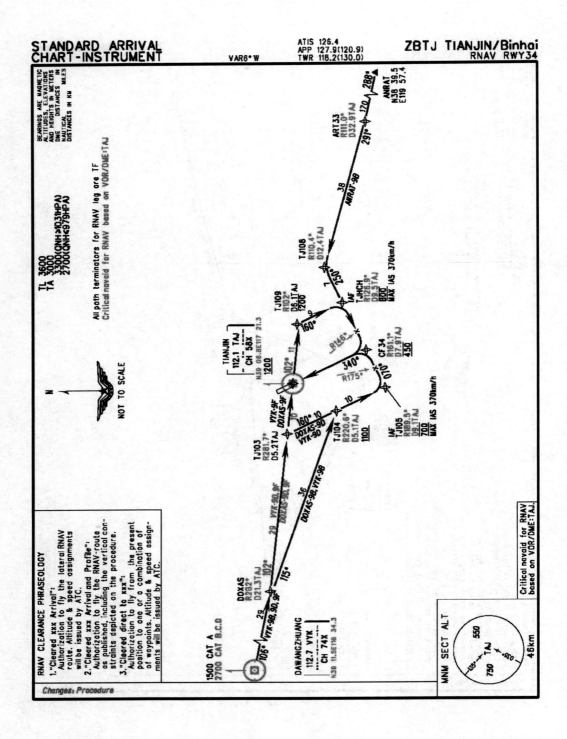

STANDARD ARRIVAL
CHART-INSTRUMENT

VAR6°W

ATIS 126.4
APP 127.9(120.9)
TWR 118.2(130.0)

ZBTJ TIANJIN/Binhai
RNAV RWY34

BEARINGS ARE MAGNETIC
ALTITUDES, ELEVATIONS
AND HEIGHTS IN METERS
DME DISTANCES IN
NAUTICAL DISTANCES MILES
DISTANCES IN KM

TL 3600
TA 3000
3300(QNH>1031HPA)
2700(QNH<979HPA)

All path terminators for RNAV leg are TF
Critical navaid for RNAV based on VOR/DME:TAJ

NOT TO SCALE

RNAV CLEARANCE PHRASEOLOGY
1. "Cleared xxx Arrival":
   Authorization to fly the lateral RNAV
   route. Altitude & speed assignments
   will be issued by ATC.
2. "Cleared xxx Arrival and Profile":
   Authorization to fly the RNAV-route
   as published, including the vertical con-
   straints depicted on the procedure.
3. "Cleared direct to xxx":
   Authorization to fly from the present
   position to one or a combination of
   waypoints. Altitude & speed assignments will be issued by ATC.

Changes: Procedure

MNM SECT ALT

Critical navaid for RNAV
based on VOR/DME:TAJ

·158·

RNAV STAR DATA　　　　　　　　　　ZBTJ TIANJIN/Binhai

Way-point list for RWY16/34

ID	COORDINATES	ID	COORDINATES
TAJ	N39 06.8　E117 21.3	CF34	N38 59.5　E117 25.7
TJ002	N39 14.6　E117 10.3	TJ108	N39 03.6　E117 36.9
CF16	N39 16.4　E117 15.3	TJ109	N39 06.1　E117 29.2
TJ004	N39 18.7　E117 21.6	TJLUX	N39 09.3　E117 27.3
TJ005	N39 13.8　E117 24.6	TJHCH	N39 01.8　E117 31.9
TJ006	N39 05.0　E117 30.0	ART33	N38 58.2　E118 02.3
TJ103	N39 07.2　E117 14.7	DOXAS	N39 08.8　E116 54.1
TJ104	N39 02.5　E117 17.7	VYK	N39 11.5　E116 34.3
TJ105	N38 57.6　E117 20.7		

Way-point sequence for RWY16

VYK9A	VYK	DOXAS	TJ002	CF16		
VYK9C	VYK	DOXAS	TAJ	TJ005	TJ004	CF16
DOXAS9A	DOXAS	TJ002	CF16			
DOXAS9C	DOXAS	TAJ	TJ005	TJ004	CF16	
ANRAT9A	ART33	TJ006	TJLUX	TJ005	TJ004	CF16

Way-point sequence for RWY34

VYK9B	VYK	DOXAS	TJ104	TJ105	CF34		
VYK9D	VYK	DOXAS	TJ103	TJ104	TJ105	CF34	
VYK9F	VYK	DOXAS	TJ103	TAJ	TJ109	TJHCH	CF34
DOXAS9B	DOXAS	TJ104	TJ105	CF34			
DOXAS9D	DOXAS	TJ103	TJ104	TJ105	CF34		
DOXAS9F	DOXAS	TJ103	TAJ	TJ109	TJHCH	CF34	
ANRAT9B	ART33	TJ108	TJHCH	CF34			

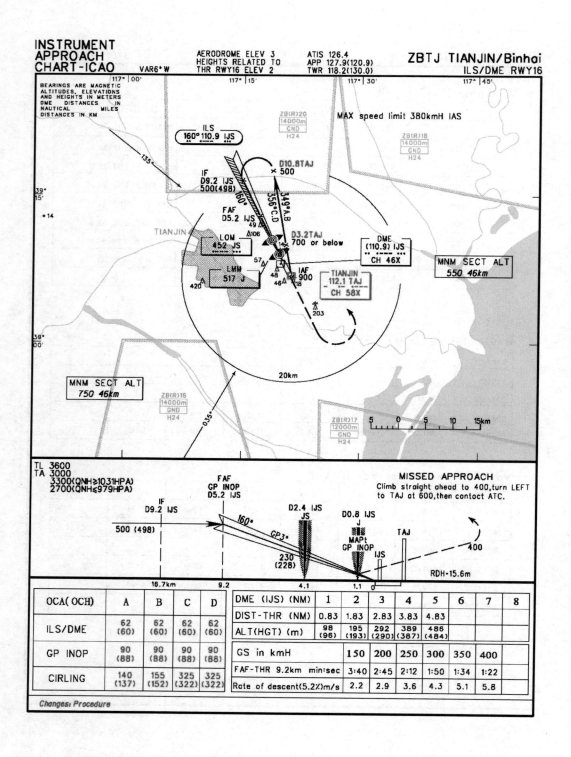

INSTRUMENT APPROACH CHART-ICAO

VAR6°W

AERODROME ELEV 3
HEIGHTS RELATED TO
THR RWY16 ELEV 2

ATIS 126.4
APP 127.9(120.9)
TWR 118.2(130.0)

ZBTJ TIANJIN/Binhai
ILS/DME RWY16

BEARINGS ARE MAGNETIC
ALTITUDES, ELEVATIONS
AND HEIGHTS IN METERS
DME DISTANCES IN
NAUTICAL MILES
DISTANCES IN KM

MAX speed limit 380kmH IAS

ILS
160° 110.9 IJS

ZB(R)20
14000m
GND
H24

ZB(R)18
14000m
GND
H24

D10.8TAJ
× 500

IF
D9.2 IJS
500(498)

349°A.B

FAF
D5.2 IJS

D3.2TAJ
700 or below

DME
(110.9) IJS
CH 46X

MNM SECT ALT
550 46km

LOM
452 JS

TIANJIN
112.1 TAJ
CH 58X

LMM
517 J

IAF
900

203

20km

MNM SECT ALT
750 46km

ZB(R)16
14000m
GND
H24

ZB(R)17
12000m
GND
H24

5   0   5   10   15km

TL 3600
TA 3000
3300(QNH≥1031HPA)
2700(QNH≤979HPA)

IF
D9.2 IJS

500 (498)

FAF
GP INOP
D5.2 IJS

160°

GP3°

230
(228)

D2.4 IJS
JS

D0.8 IJS
J

MAPt
GP INOP

IJS

TAJ

MISSED APPROACH
Climb straight ahead to 400,turn LEFT
to TAJ at 600,then contact ATC.

400

RDH-15.6m

16.7km   9.2   4.1   1.1   0

OCA(OCH)	A	B	C	D
ILS/DME	62 (60)	62 (60)	62 (60)	62 (60)
GP INOP	90 (88)	90 (88)	90 (88)	90 (88)
CIRLING	140 (137)	155 (152)	325 (322)	325 (322)

DME (IJS) (NM)	1	2	3	4	5	6	7	8
DIST-THR (NM)	0.83	1.83	2.83	3.83	4.83			
ALT(HGT) (m)	98 (96)	195 (193)	292 (290)	389 (387)	486 (484)			

GS in kmH	150	200	250	300	350	400
FAF-THR 9.2km min:sec	3:40	2:45	2:12	1:50	1:34	1:22
Rate of descent(5.2%)m/s	2.2	2.9	3.6	4.3	5.1	5.8

Changes: Procedure

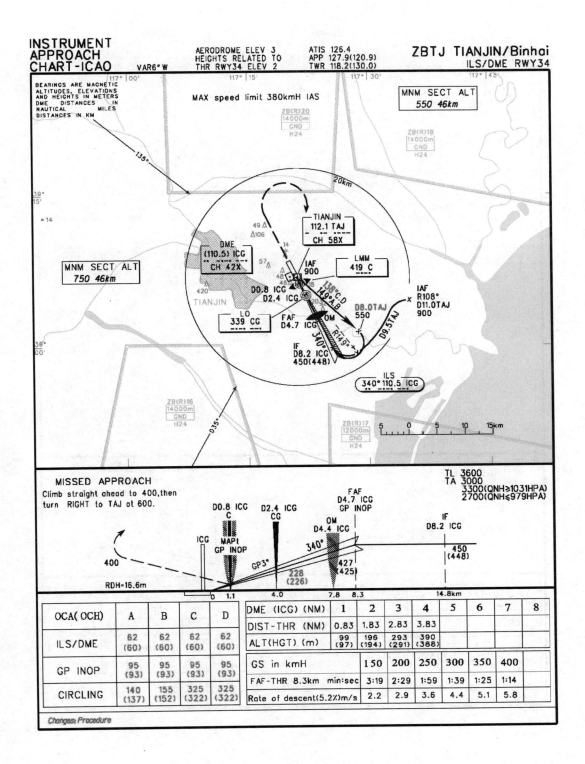

INSTRUMENT APPROACH CHART-ICAO　VAR6°W

AERODROME ELEV 3
HEIGHTS RELATED TO
THR RWY34 ELEV 2

ATIS 126.4
APP 127.9(120.9)
TWR 118.2(130.0)

ZBTJ TIANJIN/Binhai
ILS/DME RWY34

BEARINGS ARE MAGNETIC
ALTITUDES, ELEVATIONS
AND HEIGHTS IN METERS
DME DISTANCES IN
NAUTICAL MILES
DISTANCES IN KM

MAX speed limit 380kmH IAS

MNM SECT ALT
550 46km

MNM SECT ALT
750 46km

TIANJIN
112.1 TAJ
CH 58X

DME
(110.5) ICG
CH 42X

LMM
419 C

IAF
900

IAF
R108°
D11.0TAJ
900

D0.8 ICG
D2.4 ICG

138°C.D
149°A.B

D8.0TAJ
550

TIANJIN

LO
339 CG

FAF
D4.7
OM
340°

IF
D8.2 ICG
450(448)

ILS
340° 110.5 ICG

5　0　5　10　15km

MISSED APPROACH

Climb straight ahead to 400, then
turn RIGHT to TAJ at 600.

TL 3600
TA 3000
3300(QNH≥1031HPA)
2700(QNH≤979HPA)

FAF
D4.7 ICG
GP INOP

D0.8 ICG
C
D2.4 ICG
CG

OM
D4.4 ICG

IF
D8.2 ICG

ICG
MAPt
GP INOP

340°

450
(448)

400

GP3°

228
(226)

427
425)

RDH=16.6m

0　1.1　4.0　7.8　8.3　14.8km

OCA(OCH)	A	B	C	D	DME (ICG) (NM)	1	2	3	4	5	6	7	8
					DIST-THR (NM)	0.83	1.83	2.83	3.83				
ILS/DME	62 (60)	62 (60)	62 (60)	62 (60)	ALT(HGT) (m)	99 (97)	196 (194)	293 (291)	390 (388)				
GP INOP	95 (93)	95 (93)	95 (93)	95 (93)	GS in kmH		150	200	250	300	350	400	
					FAF-THR 8.3km min:sec		3:19	2:29	1:59	1:39	1:25	1:14	
CIRCLING	140 (137)	155 (152)	325 (322)	325 (322)	Rate of descent(5.2%)m/s		2.2	2.9	3.6	4.4	5.1	5.8	

Changes: Procedure

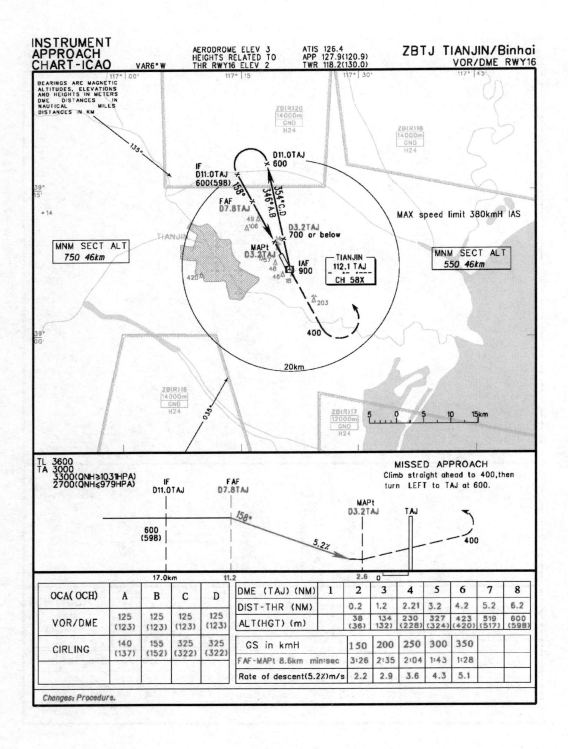

INSTRUMENT APPROACH CHART-ICAO

VAR6°W

AERODROME ELEV 3
HEIGHTS RELATED TO
THR RWY16 ELEV 2

ATIS 126.4
APP 127.9(120.9)
TWR 118.2(130.0)

ZBTJ TIANJIN/Binhai
VOR/DME RWY16

BEARINGS ARE MAGNETIC
ALTITUDES, ELEVATIONS
AND HEIGHTS IN METERS
DME DISTANCES IN
NAUTICAL MILES
DISTANCES IN KM

MAX speed limit 380kmH IAS

MNM SECT ALT
750 46km

MNM SECT ALT
550 46km

D11.0TAJ
600

IF
D11.0TAJ
600(598)

FAF
D7.8TAJ

354°C.D
346°A.B

D3.2TAJ
700 or below

MAPt
D3.2TAJ

IAF
900

TIANJIN
112.1 TAJ
CH 58X

400

20km

ZB(R)20
14000m
GND
H24

ZB(R)18
14000m
GND
H24

ZB(R)16
14000m
GND
H24

ZB(R)17
12000m
GND
H24

5 0 5 10 15km

TL 3600
TA 3000
3300(QNH≥1031HPA)
2700(QNH≤979HPA)

MISSED APPROACH
Climb straight ahead to 400,then
turn LEFT to TAJ at 600.

IF
D11.0TAJ

FAF
D7.8TAJ

MAPt
D3.2TAJ

TAJ

158°

600
(598)

5.2%

400

17.0km   11.2   2.6   0

OCA(OCH)	A	B	C	D
VOR/DME	125 (123)	125 (123)	125 (123)	125 (123)
CIRLING	140 (137)	155 (152)	325 (322)	325 (322)

DME (TAJ) (NM)	1	2	3	4	5	6	7	8
DIST-THR (NM)		0.2	1.2	2.21	3.2	4.2	5.2	6.2
ALT(HGT) (m)		38 (36)	134 (132)	230 (228)	327 (324)	423 (420)	519 (517)	600 (598)

GS in kmH	150	200	250	300	350		
FAF-MAPt 8.6km min:sec	3:26	2:35	2:04	1:43	1:28		
Rate of descent(5.2%)m/s	2.2	2.9	3.6	4.3	5.1		

Changes: Procedure.

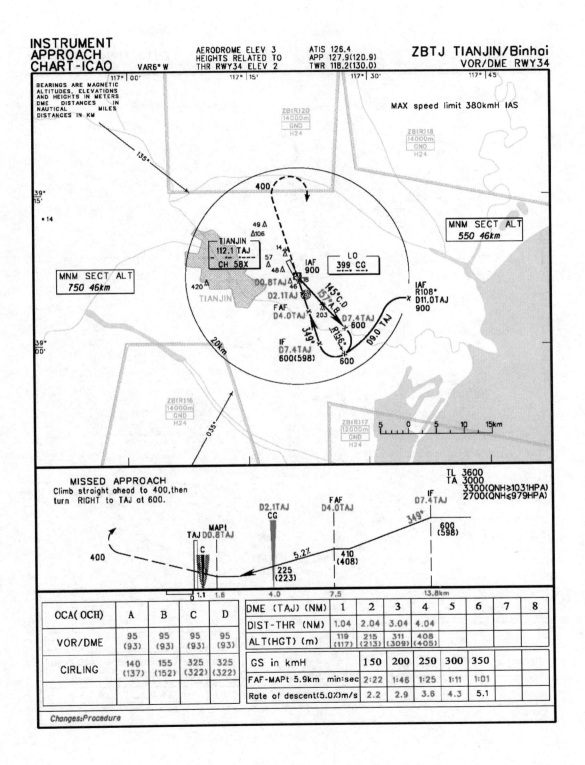

INSTRUMENT
APPROACH
CHART-ICAO   VAR6° W

AERODROME ELEV 3
HEIGHTS RELATED TO
THR RWY34 ELEV 2

ATIS 126.4
APP 127.9(120.9)
TWR 118.2(130.0)

ZBTJ TIANJIN/Binhai
VOR/DME RWY34

BEARINGS ARE MAGNETIC
ALTITUDES, ELEVATIONS
AND HEIGHTS IN METERS
DME DISTANCES IN
NAUTICAL MILES
DISTANCES IN KM

MAX speed limit 380kmH IAS

ZB(R)20
14000m
GND
H24

ZB(R)18
14000m
GND
H24

MNM SECT ALT
550 46km

TIANJIN
112.1 TAJ
CH 58X

LO
399 CG

IAF
900

IAF
R108°
D11.0TAJ
900

MNM SECT ALT
750 46km

D0.8TAJ

D2.1TAJ

FAF
D4.0TAJ

145°C.D

157A.B

D7.4TAJ
600

09.0 TAJ

IF
D7.4TAJ
600(598)

349°

R156°

600

20km

ZB(R)116
14000m
GND
H24

ZB(R)17
12000m
GND
H24

5   0   5   10   15km

MISSED APPROACH
Climb straight ahead to 400, then
turn RIGHT to TAJ at 600.

TL 3600
TA 3000
3300(QNH≥1031HPA)
2700(QNH≤979HPA)

D2.1TAJ
CG

FAF
D4.0TAJ

IF
D7.4TAJ

349°

600
(598)

MAPt
TAJ D0.8TAJ

C

5.2%

410
(408)

400

225
(223)

0 1.1 1.6   4.0   7.5   13.8km

OCA(OCH)	A	B	C	D	DME (TAJ) (NM)	1	2	3	4	5	6	7	8
VOR/DME	95 (93)	95 (93)	95 (93)	95 (93)	DIST-THR (NM)	1.04	2.04	3.04	4.04				
					ALT(HGT) (m)	119 (117)	215 (213)	311 (309)	408 (405)				
CIRLING	140 (137)	155 (152)	325 (322)	325 (322)	GS in kmH		150	200	250	300	350		
					FAF-MAPt 5.9km min:sec		2:22	1:46	1:25	1:11	1:01		
					Rate of descent(5.0%)m/s		2.2	2.9	3.6	4.3	5.1		

Changes:Procedure

· 163 ·

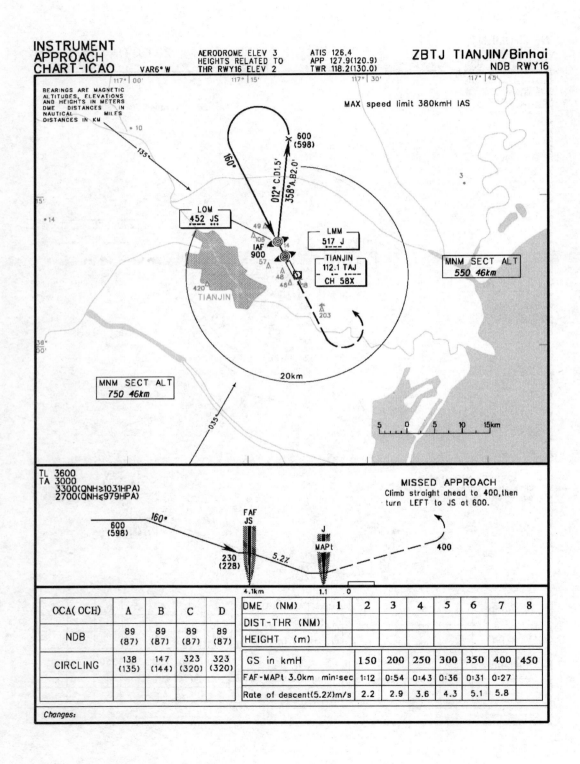

INSTRUMENT
APPROACH
CHART-ICAO   VAR6°W

AERODROME ELEV 3
HEIGHTS RELATED TO
THR RWY16 ELEV 2

ATIS 126.4
APP 127.9(120.9)
TWR 118.2(130.0)

ZBTJ TIANJIN/Binhai
NDB RWY16

BEARINGS ARE MAGNETIC
ALTITUDES, ELEVATIONS
AND HEIGHTS IN METERS
DME DISTANCES IN
NAUTICAL MILES
DISTANCES IN KM

MAX speed limit 380kmH IAS

600
(598)

012° C.D1.5'
358°A.B2.0'

160°

135

LOM
452 JS

LMM
517 J

IAF
900

TIANJIN
112.1 TAJ
CH 58X

MNM SECT ALT
550 46km

420
TIANJIN

203

MNM SECT ALT
750 46km

20km

035

5   0   5   10   15km

TL 3600
TA 3000
3300(QNH≥1031HPA)
2700(QNH≤979HPA)

MISSED APPROACH
Climb straight ahead to 400, then
turn LEFT to JS at 600.

600
(598)

160°

FAF
JS

J
MAPt

400

230
(228)

5.2%

4.1km   1.1   0

OCA(OCH)	A	B	C	D
NDB	89 (87)	89 (87)	89 (87)	89 (87)
CIRCLING	138 (135)	147 (144)	323 (320)	323 (320)

DME (NM)	1	2	3	4	5	6	7	8
DIST-THR (NM)								
HEIGHT (m)								
GS in kmH		150	200	250	300	350	400	450
FAF-MAPt 3.0km min:sec		1:12	0:54	0:43	0:36	0:31	0:27	
Rate of descent(5.2%)m/s		2.2	2.9	3.6	4.3	5.1	5.8	

Changes:

· 164 ·

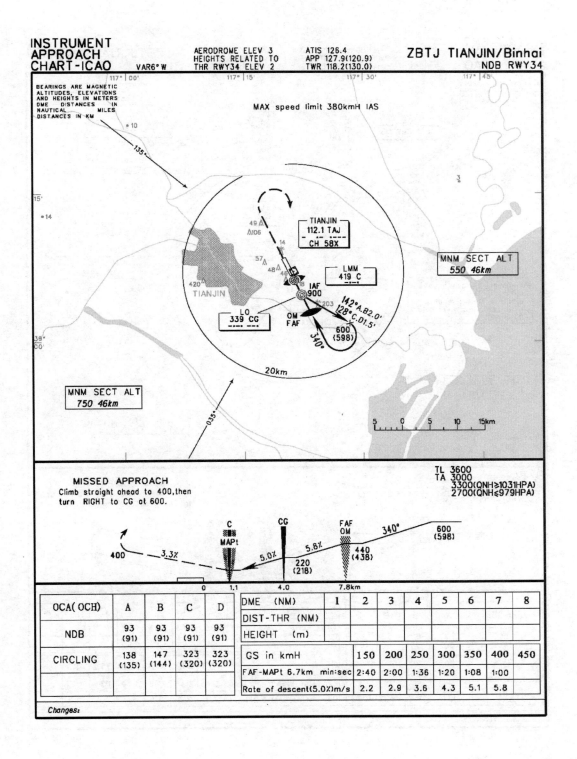

INSTRUMENT
APPROACH
CHART-ICAO   VAR6°W

AERODROME ELEV 3
HEIGHTS RELATED TO
THR RWY34 ELEV 2

ATIS 126.4
APP 127.9(120.9)
TWR 118.2(130.0)

ZBTJ TIANJIN/Binhai
NDB RWY34

BEARINGS ARE MAGNETIC
ALTITUDES, ELEVATIONS
AND HEIGHTS IN METERS
DME DISTANCES IN
NAUTICAL MILES
DISTANCES IN KM

MAX speed limit 380kmH IAS

TIANJIN
112.1 TAJ
CH 58X

LMM
419 C

MNM SECT ALT
550 46km

IAF
900

LO
339 CG

142° A.B2.0'
128° C.D1.5'

OM
FAF

600
(598)

340°

TIANJIN

20km

MNM SECT ALT
750 46km

5 0 5 10 15km

TL 3600
TA 3000
3300(QNH≥1031HPA)
2700(QNH≤979HPA)

MISSED APPROACH
Climb straight ahead to 400,then
turn RIGHT to CG at 600.

C
MAPt

CG

FAF
OM

340°

600
(598)

400   3.3%

5.0%

5.8%

440
(438)

220
(218)

0   1.1   4.0   7.8km

OCA(OCH)	A	B	C	D
NDB	93 (91)	93 (91)	93 (91)	93 (91)
CIRCLING	138 (135)	147 (144)	323 (320)	323 (320)

DME (NM)	1	2	3	4	5	6	7	8
DIST-THR (NM)								
HEIGHT (m)								
GS in kmH		150	200	250	300	350	400	450
FAF-MAPt 6.7km min:sec		2:40	2:00	1:36	1:20	1:08	1:00	
Rate of descent(5.0%)m/s		2.2	2.9	3.6	4.3	5.1	5.8	

Changes:

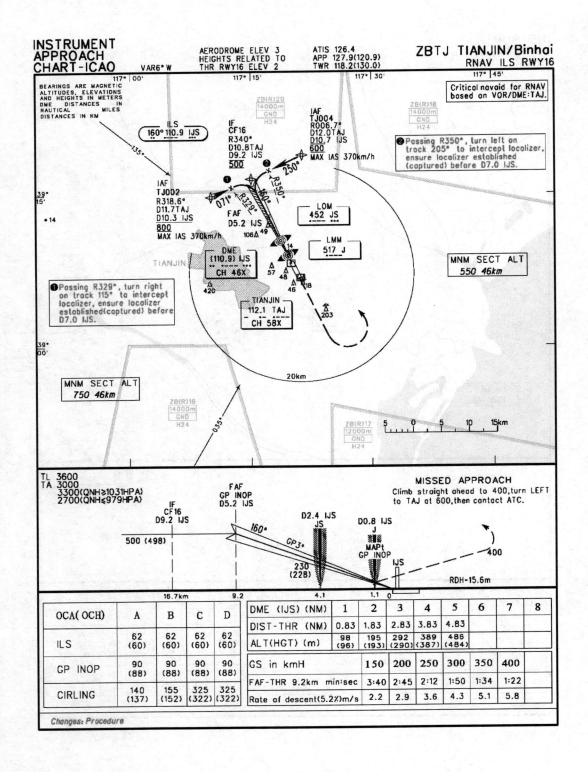

INSTRUMENT
APPROACH
CHART-ICAO

VAR6° W

AERODROME ELEV 3
HEIGHTS RELATED TO
THR RWY16 ELEV 2

ATIS 126.4
APP 127.9(120.9)
TWR 118.2(130.0)

ZBTJ TIANJIN/Binhai
RNAV ILS RWY16

BEARINGS ARE MAGNETIC
ALTITUDES, ELEVATIONS
AND HEIGHTS IN METERS
DME DISTANCES IN
NAUTICAL MILES
DISTANCES IN KM

Critical navaid for RNAV
based on VOR/DME:TAJ.

ILS
160° 110.9 IJS

IF
CF16
R340°
D10.8TAJ
D9.2 IJS
500

IAF
TJ004
R006.7°
D12.0TAJ
D10.7 IJS
600
MAX IAS 370km/h

❷ Passing R350°, turn left on
track 205° to intercept localizer,
ensure localizer established
(captured) before D7.0 IJS.

IAF
TJ002
R318.6°
D11.7TAJ
D10.3 IJS
800
MAX IAS 370km/h

FAF
D5.2 IJS

LOM
452 JS

LMM
517 J

MNM SECT ALT
550 46km

DME
(110.9) IJS
CH 46X

TIANJIN

❶ Passing R329°, turn right
on track 115° to intercept
localizer, ensure localizer
established(captured) before
D7.0 IJS.

TIANJIN
112.1 TAJ
CH 58X

20km

MNM SECT ALT
750 46km

5 0 5 10 15km

TL 3600
TA 3000
3300(QNH≥1031HPA)
2700(QNH≤979HPA)

IF
CF16
D9.2 IJS

FAF
GP INOP
D5.2 IJS

MISSED APPROACH
Climb straight ahead to 400,turn LEFT
to TAJ at 600,then contact ATC.

D2.4 IJS
JS

D0.8 IJS
J

500 (498)

160°

GP3°

MAPt
GP INOP

IJS

400

230
(228)

RDH-15.6m

16.7km   9.2   4.1   1.1   0

OCA(OCH)	A	B	C	D
ILS	62 (60)	62 (60)	62 (60)	62 (60)
GP INOP	90 (88)	90 (88)	90 (88)	90 (88)
CIRLING	140 (137)	155 (152)	325 (322)	325 (322)

DME (IJS) (NM)	1	2	3	4	5	6	7	8
DIST-THR (NM)	0.83	1.83	2.83	3.83	4.83			
ALT(HGT) (m)	98 (96)	195 (193)	292 (290)	389 (387)	486 (484)			

GS in kmH	150	200	250	300	350	400
FAF-THR 9.2km min:sec	3:40	2:45	2:12	1:50	1:34	1:22
Rate of descent(5.2%)m/s	2.2	2.9	3.6	4.3	5.1	5.8

Changes: Procedure

· 166 ·

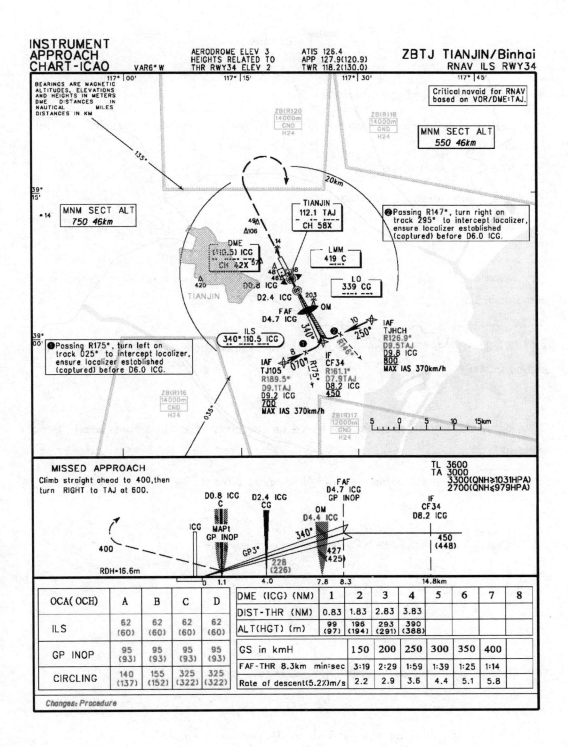

INSTRUMENT
APPROACH
CHART-ICAO

VAR6°W

AERODROME ELEV 3
HEIGHTS RELATED TO
THR RWY34 ELEV 2

ATIS 126.4
APP 127.9(120.9)
TWR 118.2(130.0)

ZBTJ TIANJIN/Binhai
RNAV ILS RWY34

BEARINGS ARE MAGNETIC
ALTITUDES, ELEVATIONS
AND HEIGHTS IN METERS
DME DISTANCES IN
NAUTICAL MILES
DISTANCES IN KM

Critical navaid for RNAV
based on VOR/DME:TAJ.

MNM SECT ALT
550 46km

ZB(R)20
14000m
GND
H24

ZB(R)118
14000m
GND
H24

135°

20km

MNM SECT ALT
750 46km

•14

TIANJIN
112.1 TAJ
CH 58X

❷Passing R147°, turn right on
track 295° to intercept localizer,
ensure localizer established
(captured) before D6.0 ICG.

49Δ
Δ106

DME
(D0.5) ICG
CH 42X

14

LMM
419 C

Δ
420

D0.8 ICG

48
48
46

57

LO
339 CG

TIANJIN

D2.4 ICG

203

OM

FAF
D4.7 ICG

340°

10

IAF
TJHCH
R126.9°
D9.5TAJ
D9.8 ICG
800
MAX IAS 370km/h

250°

ILS
340° 110.5 ICG

❶Passing R175°, turn left on
track 025° to intercept localizer,
ensure localizer established
(captured) before D6.0 ICG.

❶

8

070°

IAF
TJ105
R189.5°
D9.1TAJ
D9.2 ICG
700
MAX IAS 370km/h

R175°

R146°

❷

IF
CF34
R161.1°
D7.9TAJ
D8.2 ICG
450

ZB(R)16
14000m
GND
H24

035°

ZB(R)17
12000m
GND
H24

5   0   5   10   15km

MISSED APPROACH

Climb straight ahead to 400,then
turn RIGHT to TAJ at 600.

TL 3600
TA 3000
3300(QNH≥1031HPA)
2700(QNH≤979HPA)

FAF
D4.7 ICG
GP INOP

IF
CF34
D8.2 ICG

D0.8 ICG
C

D2.4 ICG
CG

OM
D4.4 ICG

ICG
MAPt
GP INOP

340°

450
(448)

400

GP 3°

228
(226)

427
(425)

RDH=16.6m

0 1.1   4.0   7.8 8.3   14.8km

OCA(OCH)	A	B	C	D	DME (ICG) (NM)	1	2	3	4	5	6	7	8
ILS	62 (60)	62 (60)	62 (60)	62 (60)	DIST-THR (NM)	0.83	1.83	2.83	3.83				
					ALT(HGT) (m)	99 (97)	196 (194)	293 (291)	390 (388)				
GP INOP	95 (93)	95 (93)	95 (93)	95 (93)	GS in kmH		150	200	250	300	350	400	
					FAF-THR 8.3km min:sec		3:19	2:29	1:59	1:39	1:25	1:14	
CIRCLING	140 (137)	155 (152)	325 (322)	325 (322)	Rate of descent(5.2%)m/s		2.2	2.9	3.6	4.4	5.1	5.8	

Changes: Procedure

Exercise

I. Explain the following definitions.

(1) runway physical characteristics (2) declared distances (3) seasonal availability
(4) de-icing (5) AIS briefing
(6) local traffic regulations (7) immigration

II. Answer the following questions.

(1) Which part of AIP is the aerodrome local instruction included in by a contracted state?

(2) What are the main contents for the detailed description of significant obstacles in an aerodrome AIP?

(3) What's the detailed content of approach and runway lighting system which shall be included in AIP?

III. Answer the following questions based on the given TianJin airport local instructions in 8. 2.

(1) What's the length of stopway and clearway of RWY16/34?

(2) What's the RWY16/34 runway lighting configuration?

(3) What's TL and TA of TianJin airport?

(4) What's the traffic circuit regulation of TianJin airport?

(5) What's the initial approach altitude for ILS approach runway 16?

Chapter 9

Emergency

Emergency means a Distress or an Urgency condition during the flight. The various circumstances including each emergency situation preclude the establishment of exact detailed procedures to be followed. Even though there are an infinite variety of possible emergency situations, and it's difficult for an ATS unit to prescribe a specific procedure, it's very helpful and practical to outline some general principle and guideline to handle the emergency situation. Air traffic control units must maintain full and complete coordination, and the controller must use their best judgment in handling emergency situations in case of emergency.

Pilot who encounters a distress condition declares an emergency by beginning the initial communication with the word "Mayday," preferably repeated three times. For an Urgency condition, the word "Pan-Pan" is used in the same manner. Pilot will also set the transponder to Mode Code 7700 when he/she encounters an emergency situation while Code 7500 for unlawful interference. However, if the pilot of an aircraft encountering an emergency situation has previously been directed by controller to operate the transponder on a specific code, the pilot normally maintains that code.

Even though the pilot does not use the words "Mayday" or "Pan-Pan" but the controller is in doubt that the aircraft encounter an emergency or a potential emergency situation, he/she can handle it as though it were an emergency.

When an emergency is declared by an aircraft, the ATS unit can take the following actions as below:

(a) unless clearly stated by pilot or otherwise known, take all necessary steps to ascertain aircraft identification and type, the type of emergency, the intentions of the flight crew as well as the position and level of the aircraft;

(b) decide upon the most appropriate type of assistance which can be rendered;

(c) enlist the aid of any other ATS unit or other services which may be able to provide assistance to the aircraft;

(d) provide pilot with any information requested as well as any additional relevant information, such as details on suitable aerodromes, minimum safe altitudes, weather information;

(e) obtain from the operator or the flight crew such of the following information as may be relevant: number of persons on board, amount of fuel remaining, possible presence of hazardous materials and the nature thereof; and

(f) notify relevant ATS units and authorities as specified in local instructions.

Controller has to avoid changes of radio frequency and SSR code if possible and would normally be made only when or if improved service can be provided to the aircraft concerned. He/she also issues the least manoeuvring instructions to aircraft experiencing engine failure and advises other relevant aircraft operating in the vicinity of the aircraft in emergency.

9. 1 Priority

Controller must give priority to the aircraft which is known or believed to be in a state of emergency, including being subjected to unlawful interference over other aircraft.

9. 2 Emergency handling guideline

There are some simple sets of rules to be adhered to by the controller. The emergency handling guideline by using of abbreviations seems logical, concise and is effective for controller, all the below each is offered as a possible relevant suggestion:

(a) *RISC*

Recognize that there is a problem;

Identify the relevant aircraft and arrange for special code Squawk;

Separate - Give the pilots airspace in which to operate and give them time;

Communicate with adjacent sectors/colleagues/supervisors as appropriate;

(b) *TAS*

Time - Give the pilot time to sort out the immediate problem on receipt of first notification that there is a difficulty;

Airspace - Give the pilot freedom of the adjacent airspace - get other aircraft out of the way, and off the frequency;

Silence - The controller should clear the frequency and not raise more questions than are necessary;

(c) *SSSS*

Squawk - Acknowledge the call; make sure the correct squawk is produced;

Silence - Keep the Radiotelephony (RTF) to as low a level as possible where possible assign a single frequency to the incident;

Separate - Provide appropriate and adequate airspace for the pilot to execute any essential maneuvers;

Shout - Ask for assistance from the ATC supervisor and/or colleagues;

(d) *QRST*

Quiet - Keep the frequency clear;

Recognize that there is a problem when the message is received;

Separate - Provide airspace;

Time - Give the pilot time to work on it;

(e) *ATIS*

Announcing and acknowledging the emergency or problem, getting the pilot to make the appropriate squawk;

Taq - Giving the pilot time, airspace and quiet;

Information exchange between pilot/controller and controller/controller;

Solving the problem as a team between controller/controller and controller/pilot;

(f) *ASSA*

Acknowledging the emergency or problem, getting the pilot to make the appropriate squawk;

Separate the traffic and support the pilot in so far as is possible;

Silence - Keep the RTF to a minimum; give the pilot time to think;

Advise supervisor and appropriate colleagues on other sectors;

(g) *ASSIST*

Acknowledge the call; get the squawk;

Separate the aircraft from other traffic. Give it room to manoeuvre;

Silence on the frequency. Provide separate frequency where possible - this prevents unnecessary clutter for the pilots;

Inform those who need to know and those who can help; inform others as appropriate;

Support the pilots in any way possible - Start to think of alternative routings, etc.

Time - Give the pilots time to collect their thoughts, don't harass them for information. Time produces good decisions.

Fig 9. 1 delineates the whole procedure for this emergency handling process.

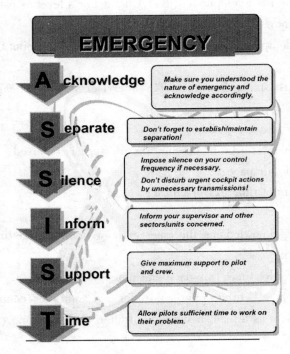

Fig 9. 1   Flow chart of emergency handling procedure and principle

9. 3   Reference emergency handling checklist for aerodrome control

It has also been proven that a handy Emergency handling checklist would be very useful and helpful for the controller to follow the handling principle and take appropriate procedure under emergency circumstances. Based on the emergency situation possible occurred in the vicinity of aerodrome and function of the aerodrome control tower, the emergency handling reference checklists for aerodrome controller pertain to possible circumstances in the vicinity of aerodrome have been given as followings:

9. 3. 1   ACAS/TCAS alerting

ACAS/TCAS is a kind of airborne conflict detection and resolution system which has been widely equipped in modern aircraft population. ACAS/TCAS alerting is generally triggered by loss of separation. ACAS/TCAS alerting when aircraft under control of aerodrome controller may result in following consequence, he/she can take appropriate measurement according to the circumstance (Table 9-1).

Table 9-1　Reference ACAS/TCAS alerting handling checklist

ACAS / TCAS

> **Expect**:	
	• Climb or descent without prior warning
	• No emergency squawk
	• Two or more aircraft involved
	• Notification from pilot of 'TCAS climb' or 'TCAS descent'
> **Remember**:	A 'Acknowledge' -S 'Separate' -S 'Silence' -I 'Inform' -S 'Support' -T 'Time'
	• When a pilot reports a manoeuvre induced by an RA: aerodrome controller never attempts to modify the aircraft flight path and provide traffic information as appropriate
	• pilots very busy
	• TCAS II altitude data is more accurate than radar data
	• Controller resumes responsibility for providing separation for all the affected aircraft when:
	Controller acknowledges a report from pilot that the aircraft has resumed the current clearance, OR
	The controller acknowledges a report from the flight crew that the aircraft is resuming the current clearance and issues an alternative clearance which is acknowledged by the flight crew
	• Following an RA event, or other significant ACAS event, pilots and controllers should complete an air traffic incident report

NB (Notice Board): Once an aircraft departs from its clearance in compliance with a RA, aerodrome controller ceases to be responsible for providing separation between that aircraft and any other aircraft affected as a direct consequence of the manoeuvre induced by the RA.

9.3.2　Birdstrike

Birdstrike is a kind of serious aviation hazard which often occurs in the vicinity of aerodrome. It may cause aircraft component damage and generally aircraft need immediate assistance (Table 9-2).

Table 9-2   Reference birdstrike handling checklist

Birdstrike	
May result in:	
	• Broken windshield / canopy
	• Engine failure (multi-engine)
	• Engine failure (single engine)
	• Hydraulic problems
	• Precautionary approach
	• Handling difficulties
	• Electrical problems
	• Gear problems
> Expect:	
	• Abandon take off
	• Immediate Landing
	• Restricted Hydraulic
> Remember:	A 'Acknowledge' - S 'Separate' -S 'Silence' - I 'Inform' -S 'Support' -T 'Time'
	• Is pilot able to control ACFT?
	• Allow long final if requested
	• Check RWY (if birdstrike after take-off)
If needed, inform pilot about:	
	• Aerodrome details as soon as possible

9.3.3   Bomb threatening

Bomb threatening is a kind of unlawful interference which is a tough task for the controller. The key to handle this problem is to expect the threatened aircraft action as soon as possible and effective coordination with the relevant authorities (Table 9-3).

Table 9-3 Reference bomb threatening handling checklist

Bomb threatening	
> Expect:	
	• ACFT may stop climb
	• Request for immediate level re-clearance
	• Landing next suitable aerodrome
	• ACFT early in landing configuration
	• RWY in use, length, surface, elevation, ILS- and NAV-frequencies
> Remember:	A 'Acknowledge' -S 'Separate' -S 'Silence' - I 'Inform' -S 'Support' -T 'Time'
	• Clear airspace in the aircraft vicinity
	• Ask for flying time needed
	• Evacuation after landing
	• Additional stairs required
	• Clear RWY according to local instructions
	• Keep safety strip clear
	• Arrange parking away from buildings/other aircraft
If needed, inform pilot about:	
	• Aerodrome details as soon as possible

9.3.4 Brake problem

Brake Problem may not affect the performance of the aircraft in-flight, but could make serious problem when landing and even affect landing airport operation. To inform the relevant department and coordination for ground preparation is necessary for this situation (Table 9-4).

Table Table 9-4 Reference brake problems handling checklist

Brake problems	
> Expect:	
	• Pilots request longest RWY
	• Overrunning RWY threshold at far end
	• Tyre burst
	• ACFT may swerve off RWY
	• RWY blocked after landing
> Remember:	A 'Acknowledge' -S 'Separate' -S 'Silence' -I 'Inform' -S 'Support' -T 'Time'
	• Inform pilot about RWY length / condition
	• Keep safety strip clear
	• Towing equipment on stand-by as appropriate
	• Technical staff required

9. 3. 5　Communication failure

Generally, the major measurement for controller to deal with the communication failure includes two aspects. One is to take all possible means establish communication with the malfunction aircraft while monitoring the aircraft maneuverings and the other is to expect the aircraft action and take appropriate measurements (Table 9-5).

Table 9-5　Reference communication failures handling checklist

Communication failure	
> **Expect**:	
When VMC	• Squawk 7600
	• Continue in VMC
	• Land at nearest suitable aerodrome
	• Report arrival by the most expeditious means to the appropriate ATS unit
When IMC	• Squawk 7600
	• The ACFT will proceed to the designated navigational aid serving the destination aerodrome and hold until commencement of descent
	• Commence descent at, or as close as possible to, the EAT last received and acknowledged, OR if no EAT received and acknowledged, commence descent at, or as close as possible to, the ETA resulting from the CPL
	• Complete a normal instrument approach procedure as specified for the designated navigation aid
	• Land, if possible, within 30 minutes after the ETA specified, OR within 30 minutes of the last acknowledged expected approach time, whichever is later
	• If being radar vectored, or proceeding offset according to RNAV without a specified limit, proceed in the most direct manner possible to rejoin the CPL no later than the next significant point, taking into consideration the applicable minimum flight altitude
> **Remember**:	A 'Acknowledge' -S 'Separate' -S 'Silence' -I 'Inform' -S 'Support' -T 'Time'
	• Failure of transmitter or receiver only
	• Possible relay by other stations?

9. 3. 6　Electrical problems

Due to loss of all generators (alternators) or battery power, electrical problems may cause a variety results to different components and systems such as communication failure, transponder switched off and even mechanic problems. Aerodrome controller deals with the problem based on clarifying the malfunction caused by the electrical problem(Table 9-6).

Table 9-6   Reference electrical problem handling checklist

Electrical problems
Loss of all generators（alternators）or battery power only power supply reduced to emergency level

> Expect:	
	• High stress level in the cockpit
	• NAV – failure, including compass
	• Transponder switched off（save energy）
	• Communication failure
	• Limited readbacks
	• Level changes to maintain VMC
	• Manual gear extension
	• Possible engine failure
> Remember:	A 'Acknowledge' -S 'Separate' -S 'Silence' -I 'Inform' -S 'Support' -T 'Time'

> If needed, inform pilot about:
• Next suitable aerodrome
• Aerodrome details as soon as possible
• Suitable vectors and position information
• Save energy
• Avoid IMC

9. 3. 7   Engine failure

In the vicinity of aerodrome, engine failure may result in take-off abortion, difficult landing and fuel dumping etc. Aerodrome controller need to give necessary assistance and offer priority for the malfunction aircraft immediately（Table 9-7）.

Table 9-7　Reference engine failure handling checklist

Engine failure	
May result in:	• Abandoned Take-off
	• Pressurization Problems
	• Fuel Dumping
	• Precautionary Approach
> Expect:	
	• Heavy workload in the cockpit
	• Deviation from SID
	• Intermediate level-off
	• Descent
	• Course deviation
	• Pressurization problems
> Remember:	A 'Acknowledge' -S 'Separate' -S 'Silence' -I 'Inform' -S 'Support' -T 'Time'
	• Inform landing aerodrome
	• Keep safety strip clear
	• Offer pilot extended final
	• Towing equipment on stand-by as appropriate
	• Clear RWY according to local instructions
	• In case of forced landing, record last known position and time
> If needed, inform pilot about:	
	• Next suitable aerodrome
	• Alternate aerodrome details as soon as possible
	• Weather information of landing aerodrome

9.3.8　Engine on fire or APU on fire

Engine on fire is very hazardous to aircraft. Just like the electrical problems, it may cause a variety results and need immediate ground assistance (Table 9-8).

Table 9-8　Reference engine on fire handling checklist

	Engine on fire or APU on fire
May result in:	• Abandoned Take-off
	• Engine Failure (multi-engine)
	• Engine Failure (single engine)
	• Smoke or Fire in the Cockpit
	• Emergency Landing
> Expect:	
	• Heavy workload in cockpit
	• Engine shutdown / fire extinguishing
	When ACFT on the Ground:
	• Hot brakes
	• Passenger evacuation
	• RWY blocked
	When ACFT in the Air:
	• Pressurization problems
	• ACFT losing altitude
	• Landing next suitable aerodrome
	• Possible diversionary or forced landing (single engine ACFT)
> Remember:	A 'Acknowledge' -S 'Separate' -S 'Silence' -I 'Inform' -S 'Support' -T 'Time'
	• Ask if dangerous goods on board
	• Ask for number of Persons On Board (POB)
	• Inform landing aerodrome
	• Clear RWY according to local instructions
	• Keep safety strip clear
	• In case of diversionary or forced landing, record last known position and time
	> If needed, inform pilot about:
	• Next suitable aerodrome
	• Aerodrome details as soon as possible
	• Weather information of landing aerodrome
	• Observed fire and/or smoke

9.3.9　Fuel problems-critical fuel status

Fuel problems generally means aircraft in serious danger and may need forced landing, so besides offering priority landing, aerodrome controller also has to inform relevant authority and prepare rescue (Table 9-9) .

Table 9-9   Reference fuel problems handling checklist

	Fuel problems – critical fuel status
May result in:	• Engine Failure (multi-engine)
	• Engine Failure (single engine)
	• Diversionary or Forced Landing
> Expect:	
	• MAYDAY: low on fuel emergency with imminent danger to ACFT
	• PAN PAN: minimum fuel ACFT needs priority handling
	• Improper use of phraseology, verify actual fuel status (low on⋯, minimum ⋯ or minimum diversion fuel)
> Remember:	A 'Acknowledge' -S 'Separate' -S 'Silence' -I 'Inform' -S 'Support' -T 'Time'
	• Keep ACFT high (save fuel)
	• Avoid ATC-caused GO AROUND
	• Inform landing aerodrome
	• Ask if dangerous goods on board
	• Ask for number of Persons On Board (POB)
	• Clear RWY according to local instructions
	• Keep safety strip clear
	• Towing equipment on standby as appropriate
	> If needed, inform pilot about:
	• Next suitable aerodrome
	• Aerodrome details as soon as possible
	• Weather information at landing aerodrome

9.3.10   Gear problems-unsafe indication / no gear

Gear Problems may be caused by some mechanical reason resulting in malfunction for gear extension. Another possibility may be some sensor trouble resulting in wrong indication, a low pass for gear inspection might be needed (Table 9-10) .

Table 9-10   Reference gear problems handling checklist

	Gear problems-unsafe indication / no gear
May result in:	• Need for External Advice （engineering）
> Expect:	
	• GO AROUND
	• Low pass for gear inspection by specialist engineering personnel
	• Manual gear extension
> Remember:	A 'Acknowledge' -S 'Separate' -S 'Silence' -I 'Inform' -S 'Support' -T 'Time'
	• Prepare for LOW PASS for visual inspection
	• Weight reduction necessary
	• Clear RWY according to local instructions
	• Keep runway safety strip clear
	• Towing equipment on stand-by as appropriate
	> If needed, inform pilot about:
	• ACFT configuration （having consulted with company if appropriate）

9.3.11   Hydraulic problems

Hydraulic problems may cause complete or partial failure of flight controls such as gear extension, brakes, flaps, nose wheel steering and difficulty in take-off or approaching and landing. Aerodrome controller has to keep communication with the pilot and learn the limited maneuverability （Table 9-11）.

Table 9-11   Reference hydraulic problems handling checklist

	Hydraulic problems
May result in:	• Fuel Dumping
	• Gear Problems
	• Brake Problems
	• Relatively High Speed
	• Approach and Landing
> Expect:	
	• Limited manoeuvrability
	• Limited flap setting

	• Limited bank angle
	• Manual gear extension, no retraction possible
	• Holding pattern for necessary checks
	• Extended final
	• Higher approach speed on final (up to 220 Kt IAS on flapless approach)
	• Limited breaking capability
	• Possible overrun
	• RWY blocked on landing
> **Remember**:	A 'Acknowledge' -S 'Separate' -S 'Silence' -I 'Inform' -S 'Support' -T 'Time'
	• Increase vertical and lateral separation
	• Ask if dangerous goods on board
	• Ask for number of Persons On Board
	• Avoid ATC-caused GO AROUND
	• Clear RWY according to local instructions
	• Keep safety strip clear
	• Towing equipment on stand-by as appropriate
> **If needed, inform pilot about**:	
	• Next suitable aerodrome
	• Aerodrome details as soon as possible
	• Weather information of landing aerodrome
	• Fire or smoke from brakes

9.3.12   Icing

Icing is not very common in the vicinity of aerodrome compared with the other emergency situation because it is formed under special meteorological condition which generally happened at high level. Icing may result in immediate change of level and communication interference (Table 9-12).

Table 9-12　Reference icing handling checklist

Icing		
> **Expect:**		
	• Immediate change of level and/or heading	
	• Limitation in rate of climb/descent	
	• Higher speed	
> **Remember:**	A 'Acknowledge' -S 'Separate' -S 'Silence' - I 'Inform' -S 'Support' -T 'Time'	
	• Avoid holding	
	• Enable continuous climb after departure	
	• Keep safety strip clear	
	• inform other ACFT, other units and MET	
> **If needed, inform pilot about:**		
	• Check anti-icing and de-icing systems	
	• Pitot heating	
	• Stall warner heating	
	• Carburettor heating	
	• Propeller heating / de-icing	
	• Wing anti-ice / de-ice	
	• Alternate air supply	
	• Windshield heating	
	• Descent with higher power setting to increase bleed air supply	
	• Higher approach/landing speed due to increase of stalling speed	

9. 3. 13　Smoke or fire in the cockpit

In the vicinity of aerodrome, smoke or fire in the cockpit will put high stress level in cockpit and generally the aircraft need emergency landing. Besides providing information and assistance to the flight crew, aerodrome controller need to follow the emergency procedure and prepared for rescue (Table 9-13) .

Table 9-13   Reference smoke or fire in the cockpit handling checklist

Smoke or fire in the cockpit	
> **Expect**:	
	• High stress level in the cockpit
	• Shortest high-speed vector to land - nearest suitable aerodrome
	• Poor RTF（because of oxygen mask）or loss of RTF
	• Passenger evacuation
	• RWY blocked
> **Remember**:	A 'Acknowledge' -S 'Separate' -S 'Silence' -I 'Inform' -S 'Support' -T 'Time'
	• Ask if dangerous goods on board
	• Ask for number of Persons On Board （POB）
	• Inform landing aerodrome
	• Offer out of wind landing if more expeditious
	• Clear RWY according to local instructions
	• Keep safety strip clear
	• RWY lighting system 100%
> **If needed, inform pilot about**:	
	• Track miles to touchdown of next suitable aerodrome
	• Availability of automatic approach low visibility procedure
	• Aerodrome details as soon as possible
	• Weather information of landing aerodrome

9.3.14   Unlawful interference

Dealing with unlawful interference is part of an aerodrome emergency procedure. Activate the emergency procedure immediately after confirming the situation （Table 9-14）.

Table 9-14　Reference unlawful interference handling checklist

Unlawful interference

> **Expect**:	
	• Squawk 7500
	• Course / level deviations
	• No or unusual replies to RTF communication
	• No compliance with given instructions
> **Remember**:	A 'Acknowledge' -S 'Separate' -S 'Silence' -I 'Inform' -S 'Support' -T 'Time'
	• Do not initiate any further RTF referring to the hijacking unless confirmed by the pilot
	• Comply with pilot's requests as far as possible
	• Transmit pertinent information without expecting a reply
	• Collect any necessary information e. g. destination aerodrome, WX situation at destination, routing, etc.

> If needed, inform pilot about:

	• Confirm squawk
	• No reply here shall NOT be taken as an indication that the squawk was set by mistake
	• Any information requested

Exercise

I. Explain the following definitions.

(1) emergency situation　　(2) RA　　(3) birdstrike　　(4) EAT

(5) course deviation　　(6) SID　　(7) pitot

II. Answer the following questions.

(1) What does the abbreviation "ASSIST" stand for?

(2) What kind of hazardous results to aircraft may be caused by electrical system prob-

lems?

(3) In case of informed by the flight crew that they decide to make an emergency landing, what shall an aerodrome controller do?

(4) In case of engine failure, if needed, what information shall aerodrome controller inform the pilot?

(5) Write down the procedure for aerodrome controller to deal with the unlawful interference.

Glossary

Part A. Words and phraseology

A

Abnormal
反常的

Not typical, usual, or regular; not normal; deviant.

Accentuate
强调，着重

To stress or emphasize; intensify.

Airbridge
廊桥

Telescopic walkway from the terminal building to the aircraft.

ADS agreement
自动相关监视协议

An ADS reporting plan which establishes the conditions of ADS data reporting (i. e. data required by the air traffic services unit and frequency of ADS reports which have to be agreed to prior to the provision of the ADS services).

ADS contract
ADS 合同

A means by which the terms of an ADS agreement will be exchanged between the ground system and the aircraft, specifying under what conditions ADS reports would be initiated, and what data would be contained in the reports.

Aerial
天线

Antenna, a metallic apparatus for sending or receiving electromagnetic waves.

Aerodrome
机场

A defined area on land or water (including any buildings, installations and equipment) intended to be used either wholly or in part for the arrival, departure and surface movement of aircraft.

Aerodrome control service
机场管制服务

Air traffic control service for aerodrome traffic.

Aerodrome control tower
机场管制塔台

An unit established to provide air traffic control service to aerodrome traffic.

Aerodrome elevation
机场标高

The elevation of the highest point of the landing area.

Aerodrome taxi circuit
机场滑行路线

The specified path of aircraft on the maneuvering area during specific wind conditions.

Aerodrome traffic
机场交通

All traffic on the maneuvering area of an aerodrome and all aircraft flying in the vicinity of an aerodrome.

Aerodrome traffic circuit
起落航线

The specified path to be flown by aircraft operating in the vicinity of an aerodrome.

Aeronautical Information Publication
航行情报汇编

A publication issued by or with the authority of a State and containing aeronautical information of a lasting character essential to air navigation.

Aeronautical ground light
地面航行灯光

Any light specially provided as an aid to air navigation, other than a light displayed on an aircraft.

Air-taxiing
空中滑行

Movement of a helicopter/VTOL above the surface of an aerodrome, normally in ground effect and at a ground speed normally less than 37 km/h (20 kt).

Air traffic control clearance
空中交通管制许可

Authorization for an aircraft to proceed under conditions specified by an air traffic control unit.

Air traffic control instruction
空中交通管制指令

Directives issued by air traffic control for the purpose of requiring a pilot to take a specific action.

Air traffic services reporting office
空中交通服务报告室

An unit established for the purpose of receiving reports concerning air traffic services and flight plans submitted before departure.

Approach control service
进近管制服务

Air traffic control service for arriving or departing controlled flights.

Apron
机坪

A defined area, on a land aerodrome, intended to accommodate aircraft for purposes of loading or unloading passengers, mail or cargo, fuelling, parking or maintenance.

Automatic dependent surveillance (ADS)
自动相关监视

A surveillance technique in which aircraft automatically provide, via a data link, data derived from on-board navigation and position fixing systems, including aircraft identification, four dimensional position and additional data as appropriate.

Automatic terminal information service (ATIS)
自动终端情报服务

The automatic provision of current, routine information to arriving and departing aircraft throughout 24 hours or a specified portion thereof.

Aerodrome Flight Information Service (AFIS)
机场飞行情报服务

A flight information service provided to aerodrome traffic.

Airborne Collision Avoidance System
机载防撞系统

An aircraft system based on SSR transponder signals which operates independently of groundbased equipment to provide advice to the pilot on potential conflicting aircraft that are equipped with SSR transponders.

Air-ground Communications
地空通信

Two-way communication between aircraft and stations or locations on the surface of the earth.

AIRPROX
危险接近

A situation in which, in the opinion of a pilot or controller, the distance between aircraft as well as their relative positions and speed have been such that the safety of the aircraft involved was or may have been compromised.

Air Traffic Service (ATS)
空中交通服务

A generic term meaning variously, flight information service, alerting service, air traffic advisory service, air traffic control service, area control service, approach control service or aerodrome control service.

Altitude
高度

The vertical distance of a level, a point or an object considered as a point, measured from mean sea level.

Area Control Centre
区域管制中心

A unit providing en-route air traffic control services.

B

Belly
机腹

Underside of a plane's fuselage.

Base turn
四转弯

A turn executed by the aircraft during the initial approach between the end of the outbound track and the beginning of the intermediate or final approach track. The tracks are not reciprocal.

Barrette
短排灯

A row of 3-5 lights, a group of lights.

Blind transmission
盲发

A transmission from one station to another station in circumstances where two-way communication cannot be established but where it is believed that the called station is able to receive the transmission.

Broadcast
播发

Transmission of information relating to air navigation that is not addressed to a specific station or stations.

C

Chock 轮挡	Triangular shaped blocks placed on either side of a wheel to prevent it rolling.
Commissary truck 供给车	A truck which brings aircraft cleaning and catering staff to aircraft awaiting turnaround. The body of the truck can be raised on hydraulically operated scissor legs up to the level of the passenger deck.
Ceiling 云底高	The height above the ground or water of the base of the lowest layer of cloud below 6000 m (20 000 ft) covering more than half the sky.
Control zone 管制地带	Controlled airspace extending upwards from the surface of the earth to a specified upper limit.
Control Area 管制区	Controlled airspace extending upwards from a specified limit above the surface of the earth.
Controlled Airspace 管制空域	An airspace of defined dimensions within which air traffic control service is provided in accordance with the airspace classification.
Cruising Level 巡航高度层	A level maintained during a significant portion of a flight.
Cumulonimbus 积雨云	An extremely dense, vertically developed clouds with a relatively hazy outline and a glaciated top extending to great heights, usually producing heavy rains, thunderstorms, or hailstorms.

D

Data link-automatic terminal information service (D-ATIS)
数字自动航站情报通波

The provision of ATIS via data link.

Decision altitude (DA) or decision height (DH)
决断高度/决断高

A specified altitude or height in the precision approach or approach with vertical guidance at which a missed approach must be initiated if the required visual reference to continue the approach has not been established.

Delineate
描述

To depict in words or gestures.

Dependent parallel approaches
相关平行进近

Simultaneous approaches to parallel or near-parallel instrument runways where radar separation minima between aircraft on adjacent extended runway centre lines are prescribed.

Depression
低压区

A region of low atmospheric pressure.

Displacement
偏移

A vector or the magnitude of a vector from the initial position to a subsequent position.

Diverge
分散

To depart from a set course.

Drift
吹雪

Snow, moved by wind.

E

Elevation
标高

The vertical distance of a point or a level, on or affixed to the surface of the earth, measured from mean sea level.

Estimated off-block time
预计撤轮挡时间

The estimated time at which the aircraft will commence movement associated with departure.

Estimated time of arrival
预达时间

For IFR flights, the time at which it is estimated that the aircraft will arrive over that designated point, defined by reference to navigation aids, from which it is intended that an instrument approach procedure will be commenced, or, if no navigation aid is associated with the aerodrome, the time at which the aircraft will arrive over the aerodrome. For VFR flights, the time is at which estimated that the aircraft will arrive over the aerodrome.

F

Facilitate
方便，便利

To make easy or easier.

Front
锋面

A zone of transition (change) between two air masses, one warm, the other cool.

Forecast
预报

A statement of expected meteorological conditions for a specified time or period, and for a specified area or portion of airspace.

Flight Information Service (FIS)
飞行情报服务

A service provided for the purpose of giving advice and information useful for the safe and efficient conduct of flights.

Flight Level
飞行高度层

A surface of constant atmospheric pressure, which is related to a specific pressure datum, 1013. 2 mb, and is separated from other such surfaces by specific pressure intervals.

Flight Plan 飞行计划	Specified information provided to air traffic services units, relative to an intended flight or portion of a flight of an aircraft. Flight Plans fall into two categories: Full Flight Plans and Abbreviated Flight Plans.
Fluctuate 波动	To vary irregularly.

G

Generic term 专业术语	The specialized words or phraseology in a field.
Ground power unit (GPU) 电源车	A truck which is used to start aircraft engines by giving power to electrical systems when the engines are not running.
Glide path 下滑道	A descent profile determined for vertical guidance during a final approach.
Ground effect 地面效应	A condition of improved performance (lift) due to the interference of the surface with the airflow pattern of the rotor system when a helicopter or other VTOL aircraft is operating near the ground.
Ground visibility 场面能见度	The visibility at an aerodrome, as reported by an accredited observer.

H

Hail 雹	Frozen rain drops which fall as small, hard balls known as hail stones.
Helicopter 直升机	An aircraft that derives its lift from blades that rotate about an approximately vertical central axis.

Holding point 等待点	A specified location, identified by visual or other means, in the vicinity of which the position of an aircraft in flight is maintained in accordance with air traffic control clearances.
Holding procedure 等待程序	A predetermined manoeuvre which keeps an aircraft within a specified airspace while awaiting further clearance.
Heading 航向	The direction in which the longitudinal axis of an aircraft is pointed, usually expressed in degrees from North (magnetic).
Height 高	The vertical distance of a level, a point, or an object considered as a point measured from a specified datum.

I

IFR Flight 仪表飞行	A flight conducted in accordance with the instrument flight rules.
Illuminate 说明	To make understandable.
Independent parallel approaches 独立平行进近	Simultaneous approaches to parallel or near-parallel instrument runways where radar separation minima between aircraft on adjacent extended runway centre lines are not prescribed.
Independent parallel departures 独立平行离场	Simultaneous departures from parallel or near-parallel instrument runways.
Institute 制定	To prescribe.
Instrument Meteorological Conditions 仪表气象条件	Meteorological conditions expressed in terms of visibility, horizontal and vertical distance from cloud, less than the minima specified for visual meteorological conditions.

L

Landing area
着陆区

That part of a movement area intended for the landing or take-off of aircraft.

Layout
布局

A design of architecture.

Level
平飞

A generic term relating to the vertical position of an aircraft in flight and meaning variously, height, altitude or flight level.

Location indicator
地名代码

A four-letter code group formulated in accordance with rules prescribed by ICAO and assigned to the location of an aeronautical fixed station.

M

Malfunction
故障

Work improperly and faultily.

Maneuvering area
机动区

That part of an aerodrome to be used for the take-off, landing and taxiing of aircraft, excluding aprons.

Mandatory
强制的

Required or commanded by authority; obligatory.

Marshaller
地面引导员

Ground crewman who marshals (directs) aircraft maneuvers to parking area.

Minimum Descent Altitude/ Height
最低下降高/高度

An altitude/height in a nonprecision or circling approach below which descent may not be made without visual reference.

Missed Approach Point (MAPt) 复飞点	The point in an instrument approach procedure at or before which the prescribed missed approach procedure must be initiated in order to ensure that the minimum obstacle clearance is not infringed.
Missed Approach Procedure 复飞程序	The procedure to be followed if the approach cannot be continued.
Moist 潮湿	Damp, humid, slightly wet.
Movement area 活动区	That part of an aerodrome to be used for the take-off, landing and taxiing of aircraft, consisting of the manoeuvring area and the apron (s).

N

Navaid 导航设备	Navigation facility.
Near-parallel runways 近距平行跑道	Non-intersecting runways whose extended centre lines have an angle of convergence/divergence of 15 degrees or less.
Normal operating zone (NOZ) 正常运行区	Airspace of defined dimensions extending to either side of an ILS localizer course and/or MLS final approach track. Only the inner half of the normal operating zone is taken into account in independent parallel approaches.
NOTAM 航行通告	Notice distributed by means of telecommunication containing information concerning the establishment, condition or change in any aeronautical facility, service, procedure or hazard, the timely knowledge of which is essential to personnel concerned with flight operations.

No transgression zone (NTZ) 非侵入区	In the context of independent parallel approaches, a corridor of airspace of defined dimensions located centrally between the two extended runway centre lines, where a penetration by an aircraft requires a controller intervention to manoeuvre any threatened aircraft on the adjacent approach.

O

Orientation 方向，方位	Direction, location or position relative to the points of the compass.

P

Passenger steps 客梯车	A truck which can raise a platform adopt telescopically to aircraft door.
Pier 连廊	Elevated access leading from the terminal to the airbridge.
Pilotage 驾驶术，（地标）领航	The technique or act of piloting.
Prescribe 规定	To set down as a rule or guide.
Precipitation 降水，降水量	Any form of water, such as rain, snow, sleet, or hail, that falls to the earth's surface; the quantity of such water falling in a specific area within a specific period.
Procedure turn 程序转弯	A manoeuvre in which a turn is made away from a designated track followed by a turn in the opposite direction to permit the aircraft to intercept and proceed along the reciprocal of the designated track.
Propeller 螺旋桨	Machine for propelling an aircraft or a boat, consisting of a power-driven shaft with radiating blades.

R

Ramp
机坪

This word is generally used for that area of the apron where servicing and maintenance are normally carried out.

Radar Approach
雷达进近

An approach, executed by an aircraft, under the direction of a radar controller.

Radar Identification
雷达识别

The process of correlating a particular radar blip or radar position symbol with a specific aircraft.

Radar Vectoring
雷达引导

Provision of navigational guidance to aircraft in the form of specific headings, based on the use of radar.

Reference code
参考代码

A method to provide a simple method for interrelating the numerous specifications concerning the characteristics of aerodromes so as to provide a series of aerodrome facilities that are suitable for aeroplanes that are intended to operate at an aerodrome.

Reporting Point
报告点

A specified geographical location in relation to which the position of an aircraft can be reported.

Rescue coordination centre
救援协调中心

An unit responsible for promoting efficient organization of search and rescue services and for coordinating the conduct of search and rescue operations within a search and rescue region.

Runway
跑道

A defined rectangular area on a land aerodrome prepared for the landing and take-off of aircraft.

Runway-holding position
跑道外等待点

A designated position intended to protect a runway, an obstacle limitation surface, or an ILS/MLS critical/sensitive area at which taxiing aircraft and vehicles shall stop and hold, unless otherwise authorized by the aerodrome control tower.

Runway strips
升降带

A defined area including the runway and stopway intended to reduce the risk of damage to aircraft running off a runway and to protect aircraft flying over it during take-off or landing operations.

Runway visual range
（RVR）
跑道视程

The range over which the pilot of an aircraft on the centre line of a runway can see the runway surface markings or the lights delineating the runway or identifying its centre line.

S

Signal Area
信号区

An area on an aerodrome used for the display of ground signals.

Special VFR Flight
特殊目视飞行

A flight made at any time in a control zone which is Class A airspace or is in any other control zone in IMC or at night, in respect of which the appropriate air traffic control unit has given permission for the flight to be made in accordance with special instructions given by that unit, instead of in accordance with the Instrument Flight Rules and in the course of which flight the aircraft complies with any instructions given by that unit and remains clear of cloud and in sight of the surface.

Straight Ahead
直飞

When used in departure clearances means: ' track extended runway centre-line'. When given in missed approach procedures means: 'continue on final approach track'.

Servicing truck
服务车

Truck used for servicing the aircraft (e. g. fuel and water tankers, waste disposal trucks. etc.)

Sequence flashers 顺序闪光灯	Lights which flash in sequence, one after another.
Slipstream 滑流	The turbulent flow of air driven backward by the propeller or propellers of an aircraft.
Stand 停机位	Parking place. The position occupied by an aircraft for embarkation.
Stripe 条	A long thin or broad line marked on a surface.
Strobe lights 闪光灯	Flashing lights.
Slush 雪水	Water-saturated snow which with a heel-and-toe slapdown motion against the ground will be displaced with a splatter.
Squall 飑线	A brief, sudden, violent windstorm, often accompanied by rain or snow.
Standard instrument arrival 标准进场程序	A designated instrument flight rule (IFR) arrival route linking a significant point, normally on an ATS route, with a point from which a published instrument approach procedure can be commenced.
Standard instrument departure 标准离场程序	A designated instrument flight rule (IFR) departure route linking the aerodrome or a specified runway of the aerodrome with a specified significant point, normally on a designated ATS route, at which the en-route phase of a flight commences.
Stopway 停止道	A defined rectangular area on the ground at the end of take-off run available prepared as a suitable area in which an aircraft can be stopped in the case of an abandoned take-off.

Symmetrical 对称的	Exact correspondence of form and constituent configuration on opposite sides of a dividing line or plane or about a center or an axis.

T

Tanker 加油车	Truck to refuel the aircraft.
Taxiway 滑行道	A defined path on a land aerodrome established for the taxiing of aircraft and intended to provide a link between one part of the aerodrome and another.
Rapid exit taxiway 快速脱离道	A taxiway connected to a runway at an acute angle and designed to allow landing aeroplanes to turn off at higher speeds than are achieved on other exit taxiways thereby minimizing runway occupancy times.
Terminal control area 终端管制区	A control area normally established at the confluence of ATS routes in the vicinity of one or more major aerodromes.
Threshold 跑道入口	The beginning of that portion of the runway usable for landing.
Touchdown 接地点	The point where the nominal glide path intercepts the runway.
Terrain 地面，地形	Ground, especially the nature or character of that ground.
Terminal 候机楼	The main passenger building of the airport.
Transponder 应答机	Radio apparatus which reacts to interrogation from signals received, and emits a response.

Tug
拖车

A truck which tows or moves a aircraft.

V

VFR Flight
目视飞行

A flight conducted in accordance with the visual flight rules.

Vicinity
附近

The state of being near in space or relationship.

W

Waterspout
水龙卷

A tornado or lesser whirlwind occurring over water and resulting in a funnel-shaped whirling column of air and spray.

Part B. Acronyms

A

AAIB	Air Accident Investigation Branch
aal	Above Aerodrome Level
ABV	Above
ACAS	Airborne Collision Avoidance System (pronounced A-kas)
ACC	Area Control Centre
ACFT	Aircraft
AD	Aerodrome
ADC	Aerodrome Control (ler)
ADI	Aerodrome Control Instrument
ADF	Automatic Direction-Finding Equipment
ADR	Advisory Route
ADS	Automatic Dependent Surveillance
ADT	Approved Departure Time
AFTN	Aeronautical Fixed Telecommunication Network
AFIS	Aerodrome Flight Information Service
AGCS	Air Ground Communication Service
AGL	Above Ground Level
AGL	Aeronautical Ground Lighting
AH	Alert Height
AIC	Aeronautical Information Circular
AIP	Aeronautical Information Publication
AIRPROX	Aircraft Proximity (replaces Airmiss/APHAZ)
AIS	Aeronautical Information Services
AMSL	Above Mean Sea Level
ANO	Air Navigation Order
APAPI	Abbreviated Precision Approach Path Indicator
APP	Approach control
ARO	ATS reporting office
ATA	Actual Time of Arrival
ATC	Air Traffic Control (in general)
ATD	Actual Time of Departure

ATIS	Automatic Terminal Information Service
ATS	Air Traffic Service
ATSU	Air Traffic Service Unit
AT-VASIS	Abbreviated T Visual Approach Slope Indicator System
ATZ	Aerodrome Traffic Zone

B

| BLW | Below |
| BKN | Broken |

C

CAA	Civil Aviation Authority
CAAC	Civil Aviation Administration of China
CAVOK *	Visibility, cloud and present weather better than prescribed values or conditions
CAT	Category
CB	Cumulonimbus
CL5B	Calvert System 5 cross bars
C/S	Callsign
CTA	Control Area
CTL	Controller
CTR	Control Zone

D

DA	Decision altitude
DAAIS	Danger Area Activity Information Service
DACS	Danger Area Crossing Service
D-ATIS	Data link-automatic terminal information service
DF	Direction Finding
DH	Decision Height
DME	Distance Measuring Equipment
DR	Dead Reckoning
D-VOLMET	Digital Meteorological information for aircraft in flight

E

EAT	Expected Approach Time
ENR	Enroute
ETA	Estimated Time of Arrival
ETD	Estimated Time of Departure

F

FAF	Final Approach Fix
FATO	Final approach and take-off
FIR	Flight Information Region
FIS	Flight Information Service
FL	Flight Level
Ft	Foot (feet)

G

GAT	General Air Traffic
GEN	General
GLONASS	Global Orbiting Navigation Satellite System
GMC	Ground Movement Control
GNSS	Global Navigation Satellite System
GPS	Global Positioning System

H

H24	Continuous day and night service
HF	High Frequency
HJ	Sunrise to Sunset
HN	Sunset to Sunrise

I

IAF	Initial Approach Fix

IAS	Indicated airspeed
ICAO	International Civil Aviation Organization
IF	Intermediate Approach Fix
IFR	Instrument Flight Rules
ILS	Instrument Landing System
IMC	Instrument Meteorological Conditions
INS	Inertial Navigation System
IRVR	Instrumented Runway Visual Range
ITHP	Intermediate Taxi-Holding Position

K

Kg	Kilogramme (s)
KHz	Kilohertz
Km	Kilometre (s)
Kt	Knot (s)

L

LDA	Landing distance available
LOA	Letter of agreement

M

MA/H	Minimum Altitude/Height
MAPt	Missed Approach Point
MATZ	Military Aerodrome Traffic Zone
MAX	Maximum
mb	Millibars
MDA/H	Minimum Descent Altitude/Height
MEDA	Military Emergency Diversion Aerodrome
MET	Meteorological or Meteorology
METAR	Routine aviation aerodrome weather report
MHz	Megahertz
MID	Middle
MNM	Minimum

MLS   Microwave Landing System

N

NATS   National Air Traffic Services
NDB   Non-Directional Radio Beacon
NM   Nautical Miles
NOTAM   Notices to Airman
NOZ   Normal Operating Zone
NTZ   No Transgression Zone

O

OAC   Oceanic Area Control Unit
OCA   Oceanic Control Area
OCA/H   Obstacle Clearance Altitude/Height
OJT   On Job Training
OVC   Overcast

P

PAOAS   Parallel approach obstacle assessment surfaces
PAPI   Precision Approach Path Indicator
PAR   Precision Approach Radar
PCN   Pavement Classification Number
PIL   Pilot
POB   Person on board

Q

QDM   Magnetic heading (zero wind)
QDR   Magnetic bearing
QFE   The observed pressure at a specified datum (usually aerodrome or runway threshold elevation) corrected for temperature
QGH   Ground interpreted letdown procedure using DF equipment
QNE   Landing altimeter setting

QNH	Altimeter sub-scale setting to obtain elevation when on the ground and indications of elevation when in the air
QTE	True Bearing

R

RA	Resolution Advisory (see TCAS)
RCC	Rescue Co-ordination Centre
RPS	Regional Pressure Setting
RTF	Radiotelephone/Radiotelephony
RTHP	Runway Taxi-Holding Position
RVR	Runway Visual Range
RWY	Runway

S

SAR	Search and Rescue
SID	Standard Instrument Departure
SIGMET	Significant information concerning en-route weather phenomena which may affect the safety of aircraft operations
SKC	Sky clear
SMGC	Surface movement guidance and control
SMR	Surface Movement Radar
SRA	Surveillance Radar Approach
SSR	Secondary Surveillance Radar
STAR	Standard Instrument Arrival
STOL	Short Take-off and Landing

T

TA	Traffic Advisory (see TCAS)
TAF	Terminal Aerodrome Forecast
TCAS	Traffic Alert and Collision Avoidance System
TDZ	Touchdown zone
TCU	Towering Cumulus
TLOF	Touchdown and lift-off

TMA	Terminal Control Area
T-VASIS	T Visual Approach Slope Indicator System
TWR	Tower

U

UAS	Upper Airspace
UHF	Ultra-High Frequency
UIR	Upper Flight Information Region
UTC	Co-ordinated Universal Time

V

VASI	Visual Approach Slope Indicator
VDF	Very High Frequency Direction-Finding Station
VFR	Visual Flight Rules
VDG	Visual Docking Guidance
VHF	Very High Frequency (30 to 300 MHz)
VIS	Visibility
VMC	Visual Meteorological Conditions
VOLMET	Meteorological information for aircraft in flight
VOR	Omnidirectional Radio Range
VORTAC	VOR and TACAN combination
VTOL	Vertical Take-off and landing

W

WX	Weather

Appendix 1　Chinese radiotelephony communication of examples in Chapter 7

7. 1　Provision of information to aircraft

Example 1：

✈ "浦东地面，南方 6213，15 号桥，请求开车，通播 B 收到。"

☆ "南方 6213，可以开车。"

Example 2：

✈ "浦东地面，南方 6213，11 号门，请求开车。"

☆ "南方 6213，开车时间 30 分，修正海压 1017。"

Example 3：

✈ "浦东地面，国际 1256，31 号门，请求开车。"

☆ "国际 1256，预计离场时间 49 分，开车时间自己掌握。"

Example 4：

✈ "浦东地面，国际 824，请求上海机场天气实况。"

☆ "国际 824，1430 上海机场实况，阴天，云底高 170 米，间有雷暴，云底高 500 米，能见度 5 公里，温度负 2 度，地面风 310 度 4 米秒，修正海压 1021。"

Example 5：

✈ "沈阳塔台，韩亚 784，五边，请求道面状况。"

☆ "韩亚 784，跑道 17 湿，部分有冰雪。着陆的 737 报告道面上有冰辙。"

Example 6：

☆ "国际 879，注意，滑行道 C 西面有障碍物。靠左侧滑行以便通过。"

✈ "看见障碍物了，靠滑行道 C 左边滑行，国际 879。"

Example 7：

☆ "美联航135，浦东塔台，延长三边，五边上有B757，看到报告。"

✘ "浦东塔台，美联航135，看到活动。"

Example 8：

☆ "美联航56，浦东塔台，Ⅰ类盲降最低天气标准有效。二类盲降因中线灯不工作不能使用。"

✘ "一类盲降标准，美联航56。"

Example 9：

☆ "南方6019，浦东塔台，进近灯不工作。"

✘ "进近灯不工作，南方6019。"

Example 10：

"浦东机场，UTC时间1500，通播H，着陆起飞跑道17，跑道17盲降进近。温度18，QNH1031，阵风230度，4米秒至8米秒。能见度4千米，机场附近报告有闪电。少云云高400米，多云云高700米。首次联系报告通播Hotel收到。"

Example 11：

✘ "浦东塔台，美联航566，请求低空通场，左起落架灯显示不正常。"

☆ "美联航566，可以低空通场跑道17，不低于100米，五边报告。"

✘ "跑道17不低于100米，五边报告，美联航566。"

✘ "浦东塔台，美联航566，五边了。"

☆ "美联航566，明白，左起落架看起来已经收上。"

Example 12：

☆ "南方6005，可以起飞，注意尾流。"

✚ "可以起飞，南方6005。"

Example 13：

★ "地面，9号卡车，请求前往维修基地。"

☆ "9号卡车，给右侧的美联航767让路，然后前往维修基地，注意发动机喷流。"

7. 2　Control of aerodrome traffic

Example 1：

✈ "浦东地面，南方6501，31号门，请求推出。"

☆ "南方6501，稍等。"

☆ "南方6501，可以推出。"

Example 2：

✈ "浦东地面，美联航428，重型机，11号机位，请求滑出。"

☆ "美联航428，滑行至35号跑道等待点，地面风"085"，7米秒，修正海压1023. "

✈ "明白，滑行至跑道35，修正海压1023，美联航428。"

Example 3：

✈ "浦东地面，美联航511，10号停机位，请求滑行。"

☆ "美联航511，经K5，T5滑行至跑道36。"

✈ "明白，滑行至跑道36，美联航511。"

☆ "美联航511，T5外等待。给向西滑至停机坪的南方757让路。"

Example 4：

✈ "浦东塔台，南方6012，请求在17号跑道上逆向滑行。"

☆ "南方6012，可以在17号跑道上逆向滑行。"

Example 5：

★ "地面，6号拖车，请求将美联航757从3号维修机库拖至25号门。"

☆ "6号拖车，可以经机坪西端拖飞机至25号门。"

Example 6：

✈ "浦东地面，美联航428，重型机，11号机位，请求滑行。"

☆ "美联航428，在航行道E等前往跑道17的救护车通过。"

Example 7：

✈ "浦东塔台，日航47，准备好起飞。"

☆ "日航47，跑道外等待。"

Example 8：

✈ "浦东地面，美联航 65，重型机，请求通过跑道 35。"

☆ "美联航 65，快速通过跑道 35，五边上的 737 距跑道入口 5 公里。"

Example 9：

✈ "浦东地面，国际 952，重型机，请求通过跑道 17。"

☆ "国际 952，快速通过跑道 17，脱离跑道报告。"

✈ "国际 952，退出跑道了。"

Example 10：

🚁 "浦东地面，G-CD，请求从机坪空中滑行到维修区。"

☆ "G-CD，经滑行道 C 空中滑行至维修区，避开停靠的 B757。"

🚁 "经滑行道 C 空中滑行至维修区，G-CD。"

7.3　Control of departing aircraft

Example 1：

☆ "进近，塔台，请求放行美联航 453。"

❀ "塔台，34 分放行美联航 453。"

Example 2：

☆ "进近，塔台，我们能改在 28 分放行美联航 453 吗?"

❀ "塔台，不能在 28 分放行美联航 453，流量控制。"

Example 3：

☆ "进近，塔台，国泰 254 预计起飞时间 45 分，请求批准。"

❀ "塔台，同意国泰 254 请求。"

Example 4：

☆ "国际 952，可以按计划航路放行至大连，航路上请求高度改变，应答机编码 3131"。

✈ "经计划航路至大连，航路上请求高度层改变，应答机编码 3131，国际 952"。

Example 5：

☆ "美联航 684，滑行至跑道 17 等待点，准备好起飞报告"。

✚ "滑行至跑道 17 等待点，美联航 684。"

✚ "美联航 684，准备好起飞。"

☆ "美联航 684，跑道外等待，五边上有飞机。"

Example 6：

✚ "日航 513，准备好起飞。"

☆ "日航 513，进跑道等待。"

✚ "进跑道等待，日航 513。"

☆ "日航 513，起飞后联系离场频率 126.4，可以起飞。"

Example 7：

✚ "浦东地面，国际 1259，证实经滑行道 D 滑行至跑道 17 等待点。"

☆ "国际 1259，正确。"

Example 8：

☆ "美联航 614，可以起飞，跑道 17。"

✚ "可以起飞，跑道 17，美联航 614。"

Example 9：

☆ "美联航 321，可以立即起飞吗?"

✚ "美联航 321，准备好起飞。"

☆ "美联航 321，立即起飞否则退出跑道。"

✚ "美联航 321 开始滑跑。"

Example 10：

☆ "国际 1425，现在位置等待，我再说一遍，取消起飞，五边上有飞机。"

✚ "国际 1425，等待。"

Example 11：

✘ "浦东塔台，南方 6234 开始滑跑。"

☆ "南方 6234，立即停止，我再说一遍，南方 6234，立即停止。"

✘ "南方 6234，正在停下。"

Example 12：

☛ "浦东塔台，G-CD，请求起飞。"

☆ "G-CD，可以起飞，起飞后右转爬升至 300 米保持，联系离场 126.4。"

☛ "起飞后右转爬升至 300 米保持，联系离场 126.4，G-CD。"

Example 13：

✘ "浦东塔台，新加坡航空 3223 起飞了。"

☆ "新加坡航空 3233，保持起飞航向，稍后通知你左转。"

7.4 Control of traffic in traffic circuit

Example 1：

✘ "浦东塔台，美联航 56，跑道南边 12 公里高度 900 米准备落地。"

☆ "美联航 56，直线进近跑道 35，地面风 340，7 米秒，修正海压 1031。"

✘ "直线进近跑道 35，修正海压 1031，美联航 56。"

Example 2：

✘ "浦东塔台，东航 126，重型机，机场北侧 10 公里，高度 700 米准备落地。通播 Bravo 收到。"

☆ "东航 126，下降至起落航线高度，加入跑道 35 右三边，修正海压 1031。"

✘ "跑道 35 右三边，修正海压 1031，东航 126"。

Example 3：

✘ "浦东塔台，南方 6019 三边。"

☆ "南方 6019，跟在五边上的 737 后面，你是第二个落地。"

✘ "跟在 737 后面第二个落地，南方 6019。"

Example 4：

✘ "浦东地面，国际 1052，三边。"

☆ "国际 1052，做小航线，三转弯报告。"

✘ "小航线，三转弯报告，国际 1052。"

7.5   Control of landing traffic

Example 1：

- ✘ "浦东塔台，南方684，长五边。"
- ☆ "南方684，跑道35可以落地，地面风270，10米秒。"

Example 2：

- ✘ "浦东塔台，国际592，请求落地连续。"
- ☆ "国际592，可以落地连续。"

Example 3：

- ✘ "浦东塔台，美联航624，请求低空通场跑道17。左起落架显示不正常。"
- ☆ "美联航624，可以低空通场跑道17，下降高度不低于50米，通场后，保持跑道航向，爬升到900米保持。"

Example 4：

- 📞 "浦东塔台，G-CD，请求直接飞向直升机5号机位。"
- ☆ "G-CD，直飞直升机5号机位，与跑道17保持距离，注意喷流，可以落地。"
- 📞 "直接飞向5号直升机停机位，与跑道17保持距离，G-CD。"

Example 5：

- ☆ "海南789，延长三边，跟在五边4公里的737后面，第二个落地。"
- ✘ "看见737了，海南789。"
- ☆ "海南789，跑道上有飞机，向右盘旋一圈，再次五边报告。"
- ✘ "向右盘旋一圈，海南789。"
- ✘ "塔台，海南789，五边了。"

Example 6：

- ✘ "塔台，南方6901，能见跑道，请求目视进近。"
- ☆ "南方6901，可以目视进近跑道18左。"
- ✘ "可以目视进近跑道18左，南方6901。"

Example 7：

- ☆ "南方6901，右转航向150，看到跑道/前方飞机报告，预计目视进近跑道18

左。"

✗ "预计目视进近跑道 18 左，看到跑道/前方航空器报告，南方 6901。"

✗ "塔台，南方 6901，看到跑道/前方航空器了。"

☆ "南方 6901，跟在 757 后面，可以目视进近跑道 18 左，注意尾流。"

Example 8：

☆ "国际 1205，预计雷达引导目视进近跑道 36 右，能见跑道报告。"

✗ "预计雷达引导目视进近跑道 36 右，能见跑道报告，国际 1205。"

Example 9：

☆ "国际 1603，前方 15 公里 777，看见报告。"

✗ "看见飞机了，国际 1603。"

☆ "国际 1603，保持目视间隔，注意尾流。"

Example 10：

☆ "南方 6013，复飞，跑道上有飞机。"

✗ "南方 6013，复飞了。"

Example 11：

☆ "国际 1029，脱离跑道后联系地面 127.5。"

✗ "地面 127.5，国际 1029。"

Example 12：

☆ "南方 6115，退出跑道，五边上有飞机，退出跑道时联系地面 127.5。"

✗ "地面，南方 6115，退出跑道了，请求滑行指令。"

☆ "南方 6115，地面，沿滑行道 D 向南至 F 交叉口，在第一个道口左转，滑行至停机位。"

Example 13：

✏ "浦东地面，G-CD，请求空中滑行至直升机停机区。"

☆ "G-CD，沿滑行道 T5、K4 空中滑行至直升机停机区，注意施工。"

✏ "沿滑行道 T5、K4 空中滑行至直升机停机区，G-CD。"

7.7 Example of a full flight operation radiotelephony communication

✈ "浦东离场，美联航 65，目的地北京，请求 ATC 放行许可。"

☆ "美联航 65，浦东离场，可以经计划航路前往北京，航路上请求改变高度层，应答机编码 5151，联系地面 124.5。"

✦ "可以经计划航路前往北京，航路上请求改变高度层，应答机编码 5151，联系地面 124.5，美联航 65，再见。"

…

✦ "地面，美联航 65 重型机，24 号门，请求推出。"

☆ "美联航 65 重型机，稍等，你后面有 737 滑行，预计等待 2 分钟。"

✦ "美联航 65 重型机，稍后推出。"

☆ "美联航 65 重型机，可以推出。"

✦ "美联航 65 重型机，推出了。"

✦ "浦东地面，美联航 65，请求开车。"

☆ "美联航 65，浦东地面，预计起飞时间 49 分，开车时间自己掌握。"

✦ "预计起飞 49 分，美联航 65。"

✦ "浦东地面，美联航 65 重型机，请求滑行。"

☆ "美联航 65，经航行道 C 航行至跑道 17 等待点。注意，航行道边有施工。"

✦ "经航行道 C 滑至跑道 17 等待点，美联航 65。"

☆ "美联航 65，原地等待，给由左向右的 747 让路。"

✦ "原地等待，美联航 65。"

☆ "美联航 65，继续滑行至跑道 17 等待点，联系塔台 118.7。"

✦ "塔台 118.7，再见，美联航 65。"

…

✦ "浦东塔台，美联航 65，准备好起飞。"

☆ "美联航 65，浦东塔台，看见五边的空客报告。"

✦ "看见空客了。美联航 65。"

☆ "美联航 65，跟在着陆的空客后面，跑道外等待。"

✦ 美联航 65 重型机照办。"

☆ "美联航 65，起飞后保持跑道航向，上升高度 600 米保持。注意尾流，可以起飞，地面风 320，3 米秒。"

✦ "美联航 65 开始滑跑。"

✖ "塔台，美联航 65，起飞了。"

☆ "美联航 65，雷达识别了。保持现在航向，上升高度 600 米，联系离场 126.7。"

✖ "上升 600 米，离场 126.7，美联航 65 重型机，再见。"

…

✖ "北京塔台，美联航 65，机场南边，高度 600 米，通播 D 收到。"

☆ "美联航 65，北京塔台，下降至起落航线高度，加入 35 跑道右三边，修正海压 1031。"

✘ "跑道 35 右三边，修正海压 1031，美联航 65。"

☆ "美联航 65，跟在五边上的 737，第二个落地，三转弯报告。"

✘ "跟在五边上的 737 第二个落地，美联航 65。"

✘ "美联航 65，三转弯，看到五边上的 737 了。"

☆ "美联航 65，可以落地，地面风 340，3 米秒。"

✘ "可以落地，美联航 65。"

☆ "美联航 65，快速退出跑道，五边上有飞机，退出报告。"

✈ "美联航 65，退出跑道了。"

☆ "美联航 65，联系地面 125.7，再见。"

…

✈ "北京地面，美联航 65，退出跑道，请求滑行指令。"

☆ "美联航 65，北京地面，沿滑行道 D 向南到 F 交叉口，左转滑行至停机位。"

Appendix 2　Relevant aircraft performance data

A300

A300　　　　　　**Airbus A300**　　　　　　**L2/JH**

MTOW: ·····················165900kg

Max Rate-of-Climb: 3800ft/minute

Service Ceiling: ···············40000ft

Length/Span: ········ 54. 08/44. 84m

Cruising Speed: ·················472Kts

Minimum App. Speed: ·············150Kts

Typical No. of Seats: ···············267

Wheel Track: ····················· 9. 6m

Remark: The performance data above is for Airbus A300-600

A310

A310   **Airbus A310**   **L2/JH**

MTOW: ·······················149960kg	**Cruising Speed**: ················· 458Kts
Max Rate-of-Climb: 4200ft/min	**Minimum App. Speed**: ···········150Kts
Service Ceiling: ··············12500m	**Typical No. of Seats**: ···············218
Length/Span: ················45. 66/43. 9m	**Wheel Track**: ····················· 9. 6m

Remark: The performance data above is for A310-300

A319

A319　　　　　　　**Airbus A319**　　　　　　　**L2/JM**

MTOW: ·······················64000kg　　**Cruising Speed:** ·················· 450Kts

Initial Rate-of-Climb:　2500ft/min　　**Minimum App. Speed:** ············140Kts

Service Ceiling: ···············11212m　　**Typical No. of Seats:** ···············124

Length/Span: ········ 33.84/34.1m　　**Wheel Track:** ····················· 7.59m

A320

A320   **Airbus A320**   **L2/JM**

MTOW: ························73500kg   **Cruising Speed:** ················· 450Kts

Initial Rate-of-Climb: >2200ft/min   **Minimum App. Speed:** ···········135Kts

Service Ceiling: ···············37000ft   **Typical No. of Seats:** ···············179

Length/Span: ············37.6/33.9m   **Wheel Track:** ···················· 7.6m

A330

A330　　　　　**Airbus A330**　　　　　**L2/JH**

MTOW: ················· 257000kg	**Cruising Speed:** ················465Kts
Initial Rate-of-Climb:　3200 ft/min	**Minimum App. Speed:** ···········140Kts
Service Ceiling: ···········12424meters	**Typical No. of Seats:** ··············295
Length/Span: ···········63. 7/16. 9m	**Wheel Track:** ····················10. 49m

Remark: The performance data above is for A330-300

A340

| A340 | Airbus A340 | L4/JH |

MTOW: ···················257000kg

Initial Rate-of-Climb: 2500 ft/min

Service Ceiling: ············12424meters

Length/Span: ···········59. 4/60. 3m

Cruising Speed: ···················466Kts

Minimum App. Speed: ············150Kts

Typical No. of Seats: ···············179

Wheel Track: ·····················10. 68m

Remark: The performance data above is for A340-200

A380

A380   **Airbus A380**   **L4/JH**

MTOW: ··················560000kg   **Cruising Speed**: ················ 561Kts
Initial Rate-of-Climb: 2500 ft/min   **Minimum App. Speed**: ············150Kts
Service Ceiling: ···········13100meters   **Typical No. of Seats**: ··············555
Length/Span: ··········· 73/79. 8m   **Wheel Track**: ···················· 14. 3m

ATR72

ATR72   **ATR72**   **L2/TM**

MTOW: ·····················21500kg	**Cruising Speed:** ·················250Kts
Initial Rate-of-Climb: 2100ft/min	**Minimum App. Speed:** ···········120Kts
Service Ceiling: ··············7620 meters	**Typical No. of Seats:** ···············70
Length/Span: ···············22. 67/24. 57m	**Wheel Track:** ····················4. 1m

B737

B73A （-100, 200）　　　　**Boeing B737**　　　　**L2/JM**
B73B （-300, 400, 500）
B73C （-600, 700, 800）

MTOW：················62800kg　　**Cruising Speed**：················430Kts
Initial Rate-of-Climb：　>3750ft/min　　**Minimum App. Speed**：···········130Kts
Service Ceiling：···········37000ft　　**Typical No. of Seats**：·············141
Length/Span：··········33. 4/28. 9m　　**Wheel Track**：···················5. 2m

Version：-100 and -200 are old versions with JD8T engines. -300 and onwards have more efficient CFM-56 engines. Figures above are for 737-300

B747

B74A （-100, 200, 300） **Boeing B747** **L4/JH**
B74B （-400）

MTOW: ························394000kg **Cruising Speed:** ················490Kts
Initial Rate-of-Climb: >1800ft/min **Minimum App. Speed:** ··········140Kts
Service Ceiling: ·············45000ft **Typical No. of Seats:** ············420
Length/Span: ···············70. 7/64. 7m **Wheel Track:** ··················11. 0m

Version: SP/-100/-100SR/-200/-300/-300SR/-400. "Freighter"/Combi versions. All
except SP have appr. the same length （SP is shorter）. Data above are for 747-400

B757

B757   **Boeing B757**   **L2/JM**

MTOW: ·················113400kg	**Cruising Speed:** ·················460Kts
Initial Rate-of-Climb: >2650ft/min	**Minimum App. Speed:** ············130Kts
Service Ceiling: ·············38000ft	**Typical No. of Seats:** ··············186
Length/Span: ··········· 47.3/38.1m	**Wheel Track:** ·················7.3m

Version: -200/-200PF (Package Freighter) / -200M (Combi).

Remark: Successor to B727, with appr. the same fuselage cross-section.

B767

B767   **Boeing B767**   **L2/JH**

MTOW: ··················181400kg	**Cruising Speed:** ··················460Kts
Initial Rate-of-Climb: >5000ft/min	**Minimum App. Speed:** ···········130Kts
Service Ceiling: ···············39700ft	**Typical No. of Seats:** ···············269
Length/Span: ···········53. 7/47. 6m	**Wheel Track:** ···················9. 3m

Version: -200, -200ER (Extended Range), -300, -300ER. Figure above are for 300ER.

Remark: Up to 3000 ft only 2000 ft/min.

B777

B777　　　　　　　**Boeing B777**　　　　　　　**L2/JH**

MTOW: ················229517kg	**Cruising Speed**: ················460Kts
Initial Rate-of-Climb:　3194ft/min	**Minimum App. Speed**: ············143Kts
Service Ceiling: ··············13137meters	**Typical No. of Seats**: ···············305
Length/Span: ················63. 73/60. 9m	**Wheel Track**: ····················11. 0m

Remark: The performance data above is for B777-200A

BAe146

BAe146   **British Arospace 146**   **L4J/M**

MTOW: ·····················42200kg	**Cruising Speed**: ·················380Kts
Initial Rate-of-Climb: >1100ft/min	**Minimum App. Speed**: ···········100Kts
Service Ceiling: ·············31000ft	**Typical No. of Seats**: ··············269
Length/Span: ·············28. 6/26. 3m	**Wheel Track**: ····················4. 7m

Version: -100, -200, -300. The latter two also available as QC (Quiet Convertible) and QT (Quiet Trader). Figure above for 146-200.

CRJ

CRJ   **Canadair**   **L2/JM**

MTOW: ⋯⋯⋯⋯⋯⋯⋯21523kg **Cruising Speed**: ⋯⋯⋯⋯⋯⋯424Kts

Initial Rate-of-Climb: 3700ft/min **Minimum App. Speed**: ⋯⋯⋯⋯135Kts

Service Ceiling: ⋯⋯⋯⋯⋯12495meters **Typical No. of Seats**: ⋯⋯⋯⋯⋯50

Length/Span: ⋯⋯⋯⋯⋯26.77/21.21m **Wheel Track**: ⋯⋯⋯⋯⋯⋯⋯3.17m

Remark: The performance data above is for CL65.

ERJ-145

EMB145　　　　　　**Embraer**　　　　　　**L2/JM**

MTOW： ·······················22100kg　　　**Cruising Speed**： ·················410Kts

Initial Rate-of-Climb： 　2379ft/min　　**Minimum App. Speed**： ···········150Kts

Service Ceiling： ·············11278meters　**Typical No. of Seats**： ··············50

Length/Span： ·············29. 87/20. 04m　**Wheel Track**： ····················4. 10m

Remark： The performance data above is for ERJ145-LR.

IL62

IL62   **Ilyushin 62**   **L4/JH**

MTOW: ·····················165000kg   **Cruising Speed:** ·················460Kts
Initial Rate-of-Climb: 3540ft/min   **Minimum App. Speed:** ············150Kts
Service Ceiling: ···············12500meters   **Typical No. of Seats:** ···············174
Length/Span: ············53. 1/12. 35m   **Wheel Track:** ····················6. 8m

Remark: The performance data above is for IL62 Mk Ⅱ.

IL76

IL76   **Ilyushin 76**   **L4/JH**

MTOW: ····················170000kg   **Cruising Speed**: ················405Kts

Initial Rate-of-Climb: 1770ft/min   **Minimum App. Speed**: ···········150Kts

Service Ceiling: ·············15500meters   **Typical No. of Seats**: ··············none

Length/Span: ··············46. 59/50. 5m   **Wheel Track**: ····················none

Version: The performance data above is for IL76T.

IL86

IL86　　　　　**Ilyushin 86**　　　　　**L4/JH**

MTOW：·····················206000kg
Initial Rate-of-Climb：　3000ft/min
Service Ceiling：···············11000meters
Length/Span：···············59. 54/48. 06m

Cruising Speed：···················484Kts
Minimum App. Speed：············153Kts
Typical No. of Seats：···············350
Wheel Track：·····················11. 15m

IL96

IL96   **Ilyushin 96**   **L4/JH**

MTOW: ·····················216000kg   **Cruising Speed:** ················460Kts

Initial Rate-of-Climb: 3000ft/min   **Minimum App. Speed:** ············150Kts

Service Ceiling: ·············39700ft   **Typical No. of Seats:** ··············300

Length/Span: ············55. 4/57. 4m   **Wheel Track:** ····················10. 4m

Remark: The performance data above is for IL96-300.

MD11

MD11   **Mcdonnell Douglas MD-11**   **L2/JH**

MTOW: ·····················285990kg   **Cruising Speed:** ·················486Kts
Initial Rate-of-Climb: 3000ft/min   **Minimum App. Speed:** ············155Kts
Service Ceiling: ··············13000meters   **Typical No. of Seats:** ···············298
Length/Span: ·············61. 2/51. 7m   **Wheel Track:** ·····················10. 57m

Remark: The performance data above is forMD-11P.

MD90

MD90   **Mcdonnell Douglas MD-90**   **L2/JM**

MTOW: ··················72803kg   **Cruising Speed:** ················456Kts

Initial Rate-of-Climb: 2800ft/min   **Minimum App. Speed:** ············150Kts

Service Ceiling: ···········11280meters   **Typical No. of Seats:** ···············153

Length/Span: ·············46. 51/32. 87m   **Wheel Track:** ··················5. 09m

Appendix 3
List of special used vehicles operating at aerodrome

Ground power unit
电源车

Mobile belt conveyor
行李传送车

Ambulance
救护车

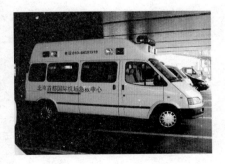

Passenger steps; Step car
客梯车

Passenger transport vehicle； Shuttle
bus; ferry bus
摆渡车

Tow bar
拖杆

Hydrant dispenser
油泵车

Fuel tanker
加油车

Fire engine；fire truck
消防车

Deicing preparation disperser
除冰液撒布车

Floor sweep vehicle
道路清扫车

De-icing cartage
除冰车

Follow-me car
引导车

Bird dispeller vehicle
驱鸟车

Surface friction tester
摩擦系数测试车

Water tanker; water service truck
清水车

Commissary truck; catering truck
航空食品车

Air starting unit; Air cart
气源车

Tug；tractor
拖车

Freight mobile platform；dolly
货物平台车

Baggage tug; transporter
行李牵引车